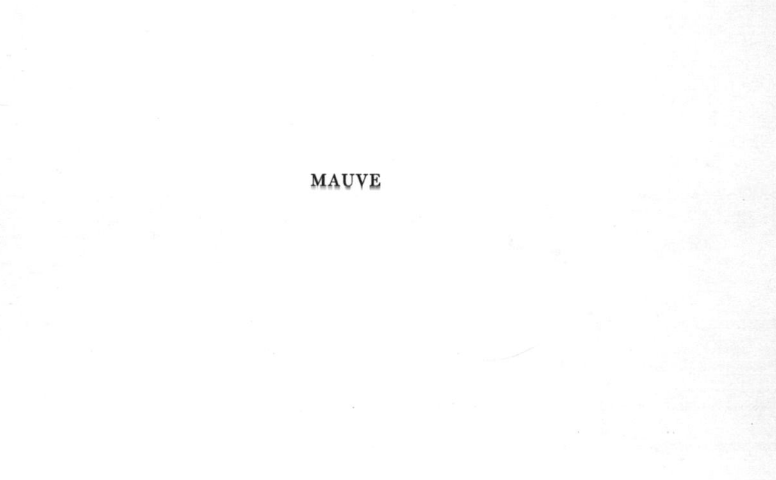

MAUVE

by the same author

EXPENSIVE HABITS: THE DARK SIDE OF THE MUSIC INDUSTRY
THE END OF INNOCENCE: BRITAIN IN THE TIME OF AIDS
THE WRESTLING
THE NATION'S FAVOURITE: THE TRUE ADVENTURES OF RADIO 1

MAUVE

How One Man Invented a Colour that Changed the World

SIMON GARFIELD

faber and faber

First published in 2000
by Faber and Faber Limited
3 Queen Square London WC1N 3AU

Typeset by Faber and Faber Limited
Printed in England by Clays Ltd, St Ives plc

A CIP record for this book
is available from the British Library

ISBN 0-571-20197-0

2 4 6 8 10 9 7 5 3 1

For BJ, Jake and Diane

CONTENTS

.. 1856 = 1906 ..

DR. W. H. PERKIN, F.R.S.

Souvenir .. of the Jubilee . of the . Coal Tar Colour Industry.

ILLUSTRATIONS

PLATE 7: The Perkin Medal (Science museum)
Perkin in 1906 (Science Museum)
Perkin's house, 'The Chestnuts' (courtesy the Perkin family)

PLATE 8: Paco Rabanne catwalk model (Chris Moore Agency)
Stained micrograph of the bacteria *Mycobacterium tuberculosis* (Science Photo Library)

Part One
INVENTION

1
THE CELEBRITY

Despite his immense wealth, Sir William Perkin seldom travelled abroad. He had visited friends and colleagues in Germany and France, and had once been to the United States, but he found the experience tiring and quickly grew weary of sightseeing. Eight days to cross the Atlantic with nothing to do but read and look at the waves. Sometimes the sea made him nauseous.

In the autumn of 1906, at the age of sixty-eight, he resolved to give travelling another chance. On 23 September he boarded RMS *Umbria*, bound for New York, taking with him his wife Alexandrine and two of his four daughters. He spent much of the voyage writing in his first-class cabin; he had a speech to give a few days after arrival, and some letters to attend to. He had recently received a request from a chemist in Germany asking for details of his early life for a lecture he hoped to deliver to his students. Perkin was famous now, and each post seemed to bring enquiries about his career and invitations to celebrations.

He wrote in a modest and unflowery style. 'The first public laboratory I worked in was the Royal College of Chemistry in Oxford Street, London, in 1853-1856.' It wasn't like the great electric laboratories of today, he noted, with your huge booming furnaces. 'There were no Bunsen burners – we had short lengths of iron tube covered with wire gauze.' It was a grey place. There were many nasty explosions.

As the *Umbria* pushed on, newspapers throughout North America excitedly carried the news of Perkin's imminent arrival. 'Famous Chemist Visits Here,' announced the *Santa Ana Evening Blade*. 'British Invade City Hall,' said the *New York Globe*. In most cities the very fact that Perkin had boarded a steamship was enough to make the front page, but the coverage was nothing compared to that greeting his arrival.

Perkin and family disembarked in New York, where they were met by Professor Charles Chandler of Columbia University. There is a photograph of them all at the quay in their heavy tweeds and woollen coats, and they don't look particularly thrilled to be there. I'm weary, Perkin told one reporter who met him at Professor Chandler's apartment in midtown Manhattan. A few days later, the *New York Herald* racked up a list of his achievements, and proclaimed: 'Coal Tar Wizard, Just Arrived in Country, Transmuted Liquid Dross To Gold'. In this story, Perkin had been elevated to the status of scientific saint, his merits placed alongside those of Watt and Stephenson, Morse and Bell.

Everyone wanted to meet him. His schedule was frantic. On Saturday night there would be a big dinner in his honour at Delmonico's, New York City's premier banqueting hall. But before then, there was some flesh-pressing and some sightseeing. On Monday he would be the guest of George F. Kunz, the gem expert at Tiffany's, who said he would escort him and his family around various stores of interest to chemists. The Perkins would then visit the zoo, New York Botanical Garden and the Museum of Art. The next day they were off to the country home, in Floyd's Neck, Long Island, of William J. Matheson, a representative of a large German chemical firm. On Wednesday he would spend time with the mayor of New York, George B. McClellan. On Thursday, H. H. Rogers would take them on his yacht for a sail up the Hudson, and the next day it would be the Laurel Hill Chemical Works. The Sunday after the banquet there would be a leisurely evening at the Chemists' Club on 55th Street.

Then there was Boston for more of the same, and then Washington DC, where Perkin was due to meet President Roosevelt. The party was then booked in at Niagara Falls, followed by Montreal and Quebec City, and then back to the United States for honorary degrees from Columbia in New York and Johns Hopkins in Baltimore.

Like many tourists before and since, Perkin found that Boston reminded him of English cities, and he especially enjoyed his trip out to Charlestown to see the battleship *Rhode Island*. 'I am greatly looking forward to meeting your President,' Perkin said as he boarded the Colonial Express bound for Washington. 'It is a certain honour,' Perkin told everyone who asked all about his great discovery. 'I was in the laboratory of the German chemist Hofmann,' he explained, his comments recorded a day later in the *Little Rock Gazette*. 'I was then eighteen. While working on an experiment, I failed, and was about to throw a certain black residue away when I thought it might be interesting. The solution of it resulted in a strangely beautiful colour. You know the rest.'

About 400 people gathered at Delmonico's at 7 p.m. One reporter present noted how 'If burial in Westminster Abbey is the highest of posthumous honours in the Anglo-Saxon world, we doubt whether a famous Englishman can receive a surer proof of his living apotheosis than when he is entertained by a company of representative Americans at Delmonico's.'

The banqueting room, a place of huge chandeliers and gilt mirrors, had been got up in English, American and German flags, and the top men (no women) from all walks of the chemical and new industrial worlds sat around forty-four tables drinking Louis Roederer Carte Blanche and telling stories about booming business and fantastic inventions. At least half of them wore fashionable moustaches. Their menu cards had been embossed, each carrying a brightly coloured tassel and a picture of Perkin looking like a benevolent country clergyman. The gold inscription read, 'Dinner in honour of Sir William Henry Perkin by his American friends to commemorate the 50th anniversary of his discovery'.

On everyone's plate lay a facsimile copy of a London patent from 1856. 'Now know ye,' it proclaimed, 'That I, the said William Henry Perkin, do hereby declare the nature of my said Invention, and in what manner the same is to be performed . . .'

Before the first course arrived, which was oysters, those disappointed with the seating arrangements took to reading the full details of Perkin's invention. The chemists among them may have been surprised at its simplicity, but they would have conceded that fifty years ago they would have been astonished.

I take a cold solution of sulphate of aniline, or a cold solution of sulphate of toluidine, or a cold solution of sulphate of xylidine, or a mixture of any one of such solutions with any others or other of them, and as much of a cold solution of a soluble bichromate as contains base enough to convert the sulphuric acid in any of the above-mentioned solutions into a neutral sulphate. I then mix the solutions and allow them to stand for ten or twelve hours, when the mixture will consist of a black powder and a solution of a neutral sulphate. I then throw this mixture upon a fine filter, and wash it with water till free from the neutral sulphate. I then dry the substance thus obtained at a temperature of 100 degrees centigrade, or 212 degrees Fahrenheit, and digest it repeatedly with coal-tar naphtha, until it is free from a brown substance which is extracted by the naphtha. I then free the residue from the naphtha by evaporation, and digest it with methylated spirit . . . which dissolves out the new colouring matter.

The men clapped and shouted Huzzah! and Hoch! as the long-bearded fellow who had composed this recipe took his seat at the top table, and began ploughing through an elaborate meal. Beyond the oysters there was clear green turtle soup. Waiters then brought radishes and olives, and Terrapin à la Maryland. The saddle of lamb Aromatic came with brussels sprouts and chestnuts, the grouse with bread sauce and currant jelly, and for dessert there was a choice of cake, cheese, coffee and Nesselrode pudding. There was more champagne. The Louis Roederer was chased by Perrier Jouet Brut and Pommery Sec. And then at about 10 o'clock it was speech time, and a small orchestra appeared at the back of the hall.

The chairman for the evening was Professor Chandler, Perkin's host in Manhattan, and he spoke of how moved he was to have such a great man in his presence. He mentioned a fund that had been set up to finance a chemical library at the Chemists Club (to be called the Perkin Library). The professor observed that there was not yet a single specialist chemistry reference library in the whole of America, and how such an institution would serve people far better than just another scholarship. He then proposed a toast to the President of the United States, the King of England and the Emperor of Germany, and everyone pushed their chairs back and joined in what they knew of 'The Star-Spangled Banner', 'Rule Britannia' and 'Die Wacht am Rhein'.

Then a man from the Mayor's office got up to read some old doggerel, which he dedicated to Perkin:

Come in the evening, or come in the morning,
Come when you're looked for, and come without warning:
A welcome and kisses you'll find here before you,
And the oftener you come the more we'll adore you.

Now it was the turn of Dr Hugo Schweitzer, a German who had worked under Robert Wilhelm Bunsen in Heidelberg. Schweitzer was also the man who had spent the best part of a year organising the present gathering. He had some alarming news: what he had to say about Perkin might take fifteen hours. The diners looked at each other, perhaps wondering what would be served for breakfast. But they cheered when Schweitzer said he hoped to condense it into fifteen minutes. A week later, one Boston newspaper would describe how, during the speech, 'vividly before one's mind . . .

trooped the great ordered cycles of the scientific progress of the last half-century'.

Schweitzer had got to know Perkin on a trip to London the previous year, and it was here that he had learnt of the background to his great discovery. 'It is hard to realise today what an epoch-making idea it was at that time,' he said. 'It was truly the spark of genius . . .'

Schweitzer explained that Perkin's discovery, which involved a specific treatment of coal-tar, was important not only for its direct and obvious effect, but also for the great many chemical advances it inspired. Perkin was indirectly responsible for enormous advances in medicine, perfumery, food, explosives and photography, and yet few beyond the immediate gathering appreciated his contribution. Even the newspapers which heralded his arrival did not fully acknowledge his achievements, and couldn't possibly estimate the debt their own trade owed to Perkin.

As Schweitzer spoke, his words were interrupted by cheers and applause. Perhaps his audience also felt envy, for it was clear that no one present could hope to match the impact that Perkin had already had upon the world. How was it that one man possessed so much energy?

In 1856, Perkin had discovered the first aniline dye, the first famous artificial colour to be derived from coal. From coal: now, fifty years later, no one regarded this as in the least bit extraordinary. But some older diners remembered the initial rumpus, the huge rage – how someone, a very young man, had found how to make colour from coal . . . If they had remembered it accurately, they would have recalled years of torment.

Now, fifty years on, there were 2,000 artificial colours, all stemming from Perkin's work. Initially, his colours were used on wool, silk, cotton and linen, but matters had progressed.

'The lady's hair is grey, or of a hue not fashionable at the time [but] coal-tar colours will assist her in appearing youthful and gay,'

Dr Schweitzer explained. 'In eating the luscious frankfurter, your soul rejoices to see the sanguineous liquid oozing from the meat – alas, coal-tar colours have done it. The product of the hen is replaced by yellow coal-tar colours in custard powders . . . leather, paper, bones, ivory, feathers, straw, grasses are all coloured, and one of the most interesting applications is the dyeing of whole pieces of furniture by dipping them in large tanks, which transforms the wood into walnut, mahogany at your command, as carried out in our big factories in Grand Rapids.'

But actually this was nothing. Perkin's discovery made sick people healthy. Coal-tar derivatives had enabled the German bacteriologist Paul Ehrlich to pioneer immunology and chemotherapy. The German scientist Robert Koch was grateful to Perkin for his discoveries of the tuberculosis and cholera bacilli. Dr Schweitzer suggested that Perkin's work had led indirectly to groundbreaking advances in the relief of pain in those with cancer.

Perhaps sensing disbelief in his audience, Schweitzer was relieved to find he could now regale them with a reasonable anecdote. He spoke of how only a few years ago a man called Fahlberg was working at Johns Hopkins and experimenting with coal-tar derivatives for scientific purposes. 'Before leaving the laboratory one evening he thoroughly washed his hands, and was under the impression that he had taken every pain in doing so. He was therefore greatly surprised on finding that, during his meal, when carrying bread to his mouth, the bread had a sweet taste.

'He suspected that his landlady had unintentionally sweetened the bread and called her to account. They had a little discussion, from which she emerged the victor. It was not the bread that tasted sweet, but his hands, and much to his surprise he noted that not only his hands but his arms had a sweet taste. The only explanation he could think of was that he had brought some chemical along from the laboratory. Rushing back to it and carefully investigating the taste of all the goblets, glasses and dishes standing on the working table, he finally came across one whose contents seemed to possess a remarkably sweet taste. Thus was made this remarkable discovery.'

Fahlberg had stumbled upon saccharin, four pounds of which possessed the sweetening power of a ton of beet sugar. He conducted some researches to find whether it was harmful to animals, and, no adverse effects being detected, was soon hailed as the founder of a huge new industry. At the time of the banquet in New York, the United States government had imposed laws banning saccharin as a sugar replacement in food on account of the devastating effects it was having on the sugar industry. This story was particularly appreciated by Professor Ira Remsen, who sat two places away from William Perkin on the top table. Fahlberg was working in Remsen's laboratory at the time of this incident.

Meanwhile, Dr Schweitzer was reaching a conclusion, and briefly mentioned that Perkin was, predictably by this stage, very much responsible for the way women smelt, having once formed coumarin from coal-tar, which led to artificial musk, and then to the artificial production of the scents of violets, roses, jasmine and the 'smell of the year' – oil of wintergreen.

The same compound which formed artificial perfume was subsequently used with nitroglycerine as an explosive in the mines and as a weapon ('the smokeless powder of the Russo-Japanese war'). Soldiers would also be grateful to Perkin for artificial salicylic and benzoic acids, both used to preserve canned foods.

At the beginning of the evening, a photographer had climbed on a ladder in the corner of the room and asked everyone to turn their chairs to face him. Almost everyone looked his way apart from Perkin, who chose to look ahead into the middle distance (Perkin was interested in the use of bags to take up the smoke of the flashlight, thus limiting the fumes of magnesium). The trick was, the photographer knew, 'I can see you if you can see me' and today we can still see them all – a remarkable record of the most distinguished chemists of the day trying their best to keep their eyes open for the duration of the long exposure.

The art of photography, naturally, was greatly enhanced by Perkin. At the time of the dinner, coal-tar preparations were responsible for the development of films and plates, and coal-tar colours improved the sensitivity of photographic emulsion, thus making it suitable for everyday snapshots. Further, in that very year, Auguste and Louis Lumière introduced Autochrome plates, the first practical application of coal-tar colour materials in photography.

Clearly, the speaker concluded, 'the world cannot spare such an extraordinary man. May his life be spared to us for many years to come, and may it be replete with health and happiness.'

This tone was sustained when Dr William Nichols, president of the US General Chemical Company, presented Perkin with the first gold impression of the Perkin Medal, henceforth to be awarded annually to only the most distinguished of American chemists. Charged with drink and the desire to better all that had gone before, Dr Nichols went for the big finish. This is the age of destruction, he announced, but his fellow chemists had a mission, and it was no less than 'saving the world from starvation'.

'Honoured by your king, by your fellow chemists, by the world,' Nichols said, as he looked down the table to Perkin, 'you may pass down the hillsides toward the setting sun with a clear conscience. You have seen the dawn of the golden age – the age of chemistry – that science which by synthesis will gather together the fragments and wastes of the other dynasties, and build for the world a civilisation which will last until the end.'

Then he sat down. A few places down the table Adolf Kutroff removed his napkin. Kutroff was one of the pioneers of the coal-tar industry in the United States, and tonight had the task of presenting Perkin with an eight-piece silver tea service, each piece inscribed with the details of one of the Englishman's discoveries.

At the very end of the dinner, and just at that time when the evening's alcohol was beginning its downward path towards stupor and headache, Sir William himself got up to speak. The crowd roused themselves once more, and really cheered. He had a deep, clear voice, and he blinked a lot as he spoke, perhaps out of modesty and shyness. Those next to him at his table noticed how he had not been drinking at all – he had been teetotal for many years. He held in his hand the speech he had written on the *Umbria*, but his first words were a mass of improvised retorts; they had thanked him, and so he must thank them, and they could have gone on back and forth like that all night. It was twenty-four years since he had last been to New York, and on his last trip far fewer people seemed to know who he was. But everything now was a great honour – the library, the medal, the tea service. 'I do not feel strange with you,' he said. 'And it may perhaps interest you to know something of my early days and how I became a chemist.'

He spoke for ten minutes about his school and his great discovery, and of the hard time he had convincing others that he had found something that might be of significance – and yet he said that even he didn't dream of what that significance might be. He was only eighteen, after all. Who else could have imagined that this filthy thick coal-tar could contain all it did? And he was lucky, because it transpired that his great invention occurred purely by chance, and it was not what he was looking for at all.

Tumult as he sat down. More toasts. Sighs as other men got up. Dr Nicholas Buller, President of Columbia, declared that democracy depends on scientific discovery. 'The age wants the man who knows. The nation will most progress that follows the advice of the men who know. The guest of the evening is a man who knows.'

Dr Ira Remsen said he knew it was getting late, but there was surely time for another rendition of 'Blessed Be the Tie that Binds.' It was a suitable song, he said. 'A pun.'

After this, the eminent scientists hailed carriages for home, or to their Manhattan hotels, and perhaps they told their partners that it had been an historical evening, and what great food. Then they all did one identical thing. Their invitation to this jubilee announced that it was a black-tie affair, but with a twist. Their dinner jackets were to be black, but their bow-ties were to be of a different colour, in recognition of the colour that had started it all off for Perkin, the colour that had chanced to change the world.

Two weeks before the event, each of the diners received a brown envelope containing a new necktie, dyed for the occasion by the St Denis Dyestuff and Chemical Company, France. The colour was often identified as a shade of purple, but for one night only there would be no mistaking its precise hue.

The men all wore it to the banquet, and now, well past midnight, they each removed it, and perhaps made a mental note to keep it safe, a perfect souvenir from a famous night in honour of a man who had invented the colour mauve.

2
NOT THE LAND OF SCIENCE

Sugar Ray Leonard slipped out of his red and black Ferrari Boxer Berlinetta, strode through the front door of Jamesons restaurant in Bethesda, Maryland, and made his way to the bar. Leonard always seems to be the handsomest man in the room, especially when someone calls his name and he flashes that dazzling smile, and on this August afternoon he looked as if he had stepped right out of the pages of *GQ*.

He wore a mauve cardigan, a light mauve shirt with the cuffs folded meticulously over the sweaters' cuffs, mauve suspenders embroidered with figures of Cupid.

'I feel great, I really do,' Leonard said.

Former World Welterweight Boxing Champion Sugar Ray Leonard profiled in *Sports Illustrated*, 1986

In May 1956, precisely one century after the discovery of mauve, a trades journal entitled *The Dyer, Textile Printer, Bleacher and Finisher* carried a warning for its subscribers. 'Readers who have thoughts of making a pilgrimage to Shadwell to see Perkin's birthplace would be well advised not to delay,' wrote the journal's editor Laurence E. Morris. 'For the site has been scheduled for redevelopment.' Once the developers moved in there was no stopping them. The site has been the subject of significant municipal improvement three times in the last four decades.

King David Lane, Upper Shadwell, is a short street containing Blue Gate School and an ugly office block, and practically nothing remains from the area in which William Perkin was born on 12 March 1838. Today's visitor finds that King David Lane has become one-way, built up with islands and bollards and signs. The road connects Cable Street – a string of council estates – to The Highway, a thundering four-lane parade of trucks and speeding Ford Mondeos. Number 3 King David Lane, where Perkin was born at home as the last of seven children, has been demolished. Like much of the East End of London, little looks the way it did before the last war.

The oldest structure is the parish church of St Paul's, a small building with an incisively tall spire. Built in 1669, the last of five London churches constructed during the Restoration, it has some famous names to its history. John Wesley preached here. Captain James Cook was an active parishioner and baptised his first son here. Jane Randolph, the mother of Thomas Jefferson, was also baptised at the church, as was William Perkin in 1838. There is a little graveyard around the church, but it is impossible to read the gravestones. In the church crypt you will find the Green Gables Montessori School.

Behind the church there is a footpath leading to many converted wharves, where those who live there can have breakfast on little terraces overlooking the Thames. Beneath them are offices for security guards and estate agents. At Shadwell Basin you may go angling and canoeing, and admire the view towards Canary Wharf and the Millennium Dome.

Forty years ago, Perkin's birthplace became A. E. Wolfe, beef and pork butcher. When that went, and the shop and rooms above it were knocked down, another new estate went up. Opposite this stands a council block called Martineau that once used to be 1 King David Fort, the house and stables the Perkins leased when William was in his teens. On one corner of this building there is a round blue plaque affixed by the Stepney Historical Trust: 'Sir William Henry Perkin, FRS, discovered the first aniline dyestuff, March

1856, while working in his home laboratory on this site, and went on to found science-based industry.' No one you meet who lives here today knows very much about him.

When he was in his twenties, William Perkin went to Leeds on business and found time to visit the house of his late grandfather, Thomas Perkin, born in 1757, of a line of Yorkshire farmers. Thomas became a leather worker, but his grandson was moved to find that he also had a rare hobby. On visiting his house at Black Thornton, near Ingleton, Perkin found a cellar containing what looked to be a laboratory. There was a still and a small smelting furnace, and various jars with grimy burnt mixtures. It was a strange stash to find in this rural community; on asking around, Perkin learnt that his grandfather had been an alchemist, and had attempted to transmute base metal into gold.

Thomas Perkin's leatherwork led him to London, where he appears to have switched trades to become a carpenter and boatbuilder. His only son, George Fowler Perkin, who was born in 1802, also became a carpenter, and a successful one. He employed twelve men, and engaged them exclusively in building the new terraced housing for the local dock workers. By today's standards, his family would be judged parvenu middle class.

Not long after his birth, William Perkin's family moved into a larger three-storey house close by, a few yards north of the High Street, a place known as King David Fort. They employed servants, and were one of the wealthiest families in the area. Their house stood out, a neighbourhood talking point. Shadwell, particularly the lower side by the docks, had some of the most wretched and crowded slums in the East End. One visitor in the early nineteenth century noted that 'thousands of useful tradesmen, artisans and mechanics inhabit, but their homes and workshops will not bear description, nor are the streets, courts, lanes and alleys by any means inviting.'

Victorian writers liked to remark on the extremes of London's poverty and wealth, virtue and iniquity. When Henry Mayhew viewed the city from a hot-air balloon in the middle of the century he was struck by the presence of mass destitution so close to the great institutions of trade, finance and empire. In Shadwell, the Perkins encountered such extremes on a daily basis. Disease was all around them. William Perkin was to lose both eldest sister and brother to tuberculosis. Their mother Sarah, a woman of Scottish descent who had moved to east London when she was a child, was thought never to have recovered from her losses.

The Perkins grew up opposite the police station, from where they witnessed an endless stream of the drunk and lawless. Much of the police work centred on a pub named Paddy's Goose, where local seamen sought prostitutes, and the unwary were press-ganged into the Royal Navy.

William Perkin attended the private Arbour Terrace School in Commercial Road, a few hundreds yards from his home. He was a gifted student, with many interests outside the standard curriculum. 'He showed remarkable dexterity in all kinds of hobbies,' his nephew Arthur H. Waters recalled. Waters's mother was about two years older than Perkin. 'They were fond of taking long rambles together, and William was particularly keen on natural history and botany. On one occasion he produced a large pipe and tobacco and proceeded to puff away manfully. But after a time he became so confoundedly ill that his sister had some difficulty in getting him home. William's craze for probing into everything, especially small things, seems to show that his wonderful instinct for research was present at a very early age.'

He became interested in photography when he was twelve, and at fourteen he took his own picture: he has a stony look, his broad forehead and strong features framed by dense black hair. He is done up in evening gear, or perhaps his church best, and he looks about twenty.

'I do not quite know where to begin,' he wrote to his colleague Heinrich Caro in 1891. 'But as the circumstances connected with my childhood and youth had, I believe, a good deal to do in influencing me in respect to practical matters, I have ventured to relate a little connected with that period for your private information.'

Caro, from 1869 to 1890 the principal investor at BASF, had written to Perkin a few weeks earlier, explaining that he was preparing the first major history of the dye industry and would like more details about his early life. Perkin wrote that he would help him with the facts as he could remember them, but midway through his reply he had a change of heart. 'I have now written you out an account of my early days, which I have never done before, and now I have done so feel some hesitation in sending it to you.' Why this should be so he did not say, but he remained a meek and demure man throughout his life. He said later that he believed only his work was important.

At the beginning he had no idea what he wanted to do with his life, though he fancied something artistic, or something practical he could do with his hands. 'Being interested in what I saw going on around me, I thought I would follow in my father's footsteps,' he wrote. He built wooden models, among other things of the steam trains he saw passing near his home. He was also drawn to engineering, and liked the illustrations of levers and pulleys he saw in a book called *The Artisan*. Published in 1828, this contained a popular summary of what was then known about mechanics, optics, magnetism and pneumatics, all of it written with an element of wonder and disbelief that science was moving so rapidly.

But Perkin was being pulled in other ways. 'I took a great interest in painting,' he explained, 'and for a short time had the mad idea that I should like to be an artist.' There was also music – he learnt the violin and double bass, and he and his brother and two sisters entertained thoughts of becoming a travelling quartet. But just before his thirteenth birthday a friend showed him some elementary experiments with crystals that he regarded as 'quite marvellous'.

'I saw chemistry was something far higher than any other subject that had come before me,' he remembered. 'I thought that if I could be articled to a pharmacist I should be happy.'

In another telling of the story, Perkin again flattened the drama. 'The possibility also of making new discoveries impressed me very much. I determined if possible to accumulate bottles of chemicals and make experiments.'

When he was thirteen he joined 600 other boys at the City of London School in a narrow street by Cheapside, not far from St Paul's. It was a strict institution with painful punishments for misbehaviour, but its educational outlook was progressive. On his arrival, Perkin was delighted to learn that it was one of the few schools in the country to offer lessons in chemistry, a subject believed to have little practical use (and certainly less than Latin or Greek). The course was taught twice a week in the lunch-hour by a writing master called Thomas Hall, and Perkin persuaded his father to pay an extra seven shillings each term for the privilege. He skipped lunch to attend. 'Thomas Hall noticed that I took a great interest in the lectures, and made me one of his helpers to prepare his lecture experiments. This was a wonderful lift for me . . . to work in the dismal place that was called a laboratory in that school.'

Hall suggested that Perkin might like to conduct some of the safer experiments at home, and helped him buy some glassware. Perkin's father again agreed to pay for his son's enthusiasms, although he made it clear he wished him to become an architect. Chemistry was fascinating, but there was no money in it.

Outside school, Perkin attended chemistry talks given by Henry Letheby at the London Hospital in Whitechapel Road, and both Letheby and Thomas Hall suggested to Perkin that he write to their friend Michael Faraday requesting permission to attend his monumental lectures at the Royal Institution. Faraday replied in his own hand, an act that delighted Perkin greatly, and so it was that on Saturday afternoons a fourteen-year-old boy found himself the youngest spectator of the latest developments in the peculiar science of electricity.

A few years before, the leading German scientist Justus Liebig had had some damning news for the delegates to the British Association meeting in Liverpool in 1837. 'England is not the land of science,' he declared. 'There is only widespread dilettantism, their chemists are ashamed to be known by that name because it has been assumed by the apothecaries, who are despised.'

In contrast, Liebig's teaching laboratory at the University of Giessen was the envy of all experimental chemists, and men travelled hundreds of miles to engage in what their own countries believed to be an unrewarding pursuit. There were chairs in chemistry at both Oxford and Cambridge, but the idea that the subject should be taught and learnt in the laboratory was unheard of; students were merely taught chemical history as part of a wider science course. At the University of Glasgow, a man named Thomas Thomson was probably the first to open up his laboratory to his students for practical instruction, and Thomas Graham, singled out by Liebig as a rare example of a forward-looking scientist, did the same at the city's Andersonian Institution in 1830. At the time of Perkin's birth there was no college anywhere in the country dedicated to the study of chemistry.

Liebig was an inspirational speaker, and it was his British lecture tour in the early 1840s which convinced men of influence that London needed a specialist chemical school (Liebig met the Prime Minister, Sir Robert Peel, who expressed personal interest due to his family's involvement in calico printing). There were plans to establish the Davy College of Practical Chemistry within the Royal Institution, but when these foundered, Sir James Clark, the

Queen's physician, Michael Faraday and the Prince Consort, for years a keen sponsor of scientific research, established a private subscription to finance the Royal College of Chemistry, raising some £5,000, and counting both Peel and Gladstone among its contributors.

The College opened temporary laboratories just off Hanover Square in 1845, and moved a year later to a permanent site at the south side of Oxford Street. The building was soon full with twenty-six students, and its size dictated that lectures be held at the Museum of Practical Geology in Jermyn Street. It was here that Perkin's teacher Thomas Hall first came to hear the young director of the college, August Wilhelm von Hofmann.

Hofmann was born in Giessen in 1818, and first studied mathematics and physics before taking chemistry with Liebig. His appointment at the Royal College was widely favoured by Prince Albert, not least because he believed that Hofmann would make advances directly beneficial to agriculture. And there was another reason: in the summer of 1845, the Queen and Prince Albert visited Bonn for the unveiling of the monument to Beethoven. Queen Victoria noted the occasion in her diary, and recorded what happened afterwards. 'We drove with the King and Queen [of Prussia] to Albert's former little house. It was such a pleasure for me to be able to see this house. We went all over it, and it is just as it was . . .' The lack of alteration was down to August Hofmann, who now lived there and occasionally conducted small chemical experiments in one of the rooms.

Thomas Hall believed that Perkin should enrol at the Royal College at fifteen, but there was severe opposition from Perkin's father. Why couldn't William be more like his older brother Thomas? Thomas was training to be an architect. 'My father was disappointed,' Perkin recorded years later. But Hall persuaded his father to meet Hofmann, who may have beguiled him with the exotic possibilities of benzene and aniline.

'He had several interviews with my father,' Perkin noted. 'And the end of it was that I went to study chemistry under Dr Hofmann.'

That was in 1853. Within five years, Perkin had made his fortune.

3
FLOATING IN THE AIR

Wandered in the town, to the Museum and Zoo . . . Reconstructions of Hausa and Sanghay villages – combination of indigo and pale calabash. Hunchback boy with staff and bowl and mauve purple jumper stretched like a landscape over his totally deformed body . . . A restaurant in a garden. I drank a beer on a red spotted cloth-covered table. Mosquitoes bit the hard parts of my fingers.

Bruce Chatwin in Niger, 1971, from *Photographs and Notebooks*

When Perkin left the Royal College every evening and walked along Oxford Street, his journey was illuminated by gas light. London was ablaze with gas: houses, factories and streets had been lit this way since the beginning of the century, and Perkin's laboratory work had begun to rely on gas for other fiery uses.

But this demand brought some terrible problems. Gas derived from the distillation of coal, and millions of tons were processed each year to meet demand. The process – which involved the highly combustible method of heating coal in closed vessels without oxygen – also yielded several useless and dangerous by-products: foul-smelling water, various sulphur compounds and a large amount of oily tar.

For many years these were regarded as waste; the problem was not how to utilise them but how to get rid of them. The sulphur was found to be removable with lime and sawdust, while the gas-water and tar were abandoned in streams, where they poisoned the water and killed the fish. Anyone who requested any of these by-products was given them without charge in huge barrels. Some hopeless experiments were conducted with them, and then they were again thrown away into streams. But gradually, in the years leading up to Perkin's birth, new uses were uncovered.

The gas-water was found to be rich in ammonia, and the sulphur compounds would be used in the manufacture of sulphuric acid. In Glasgow in the 1820s, Charles Macintosh found a use for the coal tar, developing a method of waterproofing cloth. He used it to prepare a special solution of rubber, applied it to two pieces of coat fabric, and called it a raincoat, but other people soon began calling it a macintosh. It was also used as a protective coating on timber, and was widely employed on the new railway system. Its combination with creosote also afforded a thick coating for wood and metals, and it was used as a disinfectant in sewage. Some patents from the 1840s even suggested the early use of tar and coal-tar pitch on road surfaces.

At the opening of the Royal College of Chemistry, coal-tar was already recognised as an immensely complex material. The first students understood that it consisted of the elements carbon, oxygen, hydrogen, nitrogen and a little sulphur, and they knew that from these combinations an inviting list of substances could be formed.

The study of modern chemistry was still in its infancy – it was only in 1788 that Antoine Lavoisier demonstrated that air was a mixture of gases which he called oxygen and nitrogen – and important advances were being made every year. Molecules such as the solvent naphtha had already been isolated in coal-tar in the 1820s, but the great challenge was now to reveal its constituent atoms, and to show how these may be modified to form other compounds. Naphtha was found to contain benzene, and, by a painstaking process of fractional distillation, this in turn was found to contain such materials as toluidine and aniline. The chemists often knew the atomic combination of each molecule – how many elements of carbon, how many of oxygen or hydrogen – but not how they fitted together. Their precise chains and points of attachment – those knobbly bead-and-metal constructions that (in the days before three-dimensional computer software) proud chemists liked to pose beside for photographs – would not be fully understood for several decades.

The research students at the Royal College thus conducted much of the exploratory work without map or compass, and some paid the price. Charles Mansfield, one of Hofmann's most enterprising students, had been discouraged from setting up dangerous large-scale coal-tar experiments at the Royal College, and yet persevered with his project in a building near King's Cross railway station. While preparing large quantities of benzene for an international exhibition in 1855, a fire broke out which consumed both him and his assistant.

It was aniline that most fascinated Professor Hofmann. He had spent much of his laboratory time in Germany investigating its possibilities, and continued his researches in Oxford Street. Crucially, he managed to impart this enthusiasm to his students.

'As a teacher he was singularly interesting and lucid,' Lord Playfair explained in a memorial lecture given in Hofmann's honour in 1893, the year after his death. 'He marshalled his arguments with great care, and as he brought them towards the conclusion, he increased in his persuasiveness and seemed to each individual student to take him into his special confidence.'

Frederick Abel, the joint-inventor of cordite, once asked himself, 'Who would not work, and even slave, for Hofmann?' Before he tackled explosives, Abel conducted an analysis of the mineral waters of Cheltenham and researched the effects of various substances on aniline (one of which was the poisonous gas cyanogen, from which his eyes suffered permanent damage). Another of his students established the composition of the air on Mont Blanc.

Strangely for a chemist, Hofmann was a rather clumsy man, once explaining to Abel that when he was younger he could hardly handle a test tube without 'scrunching' it. 'There was an indescribable charm in working with Hofmann,' Abel remembered, 'in watching his delight at a new result or his pathetic momentary depression when failure attended the attempt to attain a result which theory indicated. "Another dream is gone," he would mutter plaintively, with a deep sigh.'

One of Hofmann's principal talents appeared to be choosing the right student for the right job, and in selecting a huge variety of avenues for research. In his first five years, some thirty-six different projects were undertaken. The Queen and the Prince Consort were frequent visitors to his laboratories, and Hofmann delivered several lectures at Windsor Castle. At the Royal Institution in 1865, Hofmann delighted Prince Albert and other notables with a demonstration involving croquet balls and rods. The royals may have enjoyed his quaint English literal translations from German idioms; they were certainly interested in his students' work on soil and plants – in fact, they were keen on anything which might lead to practical applications.

William Perkin noted how his mentor used to tour the laboratories several times a day and talk to his students as if each piece of their work was of phenomenal importance. Occasionally their work did indeed carry genuine significance; most often it was mundane and doomed. And almost all the time Hofmann seemed to have done it before by himself. 'I well remember one day,' Perkin said, 'when the work was going on very satisfactorily and several new products had been obtained, he came up and commenced examining a product of the nitration of phenol one of the students had obtained by steam distillation. Taking a little of the substance in a watch glass, he treated it with caustic alkali, and at once obtained a beautiful scarlet salt. Looking up at us in his characteristic and enthusiastic way, he at once exclaimed, Gentlemen, new bodies are floating in the air!'*

Another tour was less fruitful. Once, Hofmann was holding a glass bottle containing a little water, and invited a student to pour sulphuric acid into it. The heat cracked the glass, and the acid splashed from the floor into Hofmann's eyes. 'Hofmann was sent home in a cab,' Perkin remembered, 'and had to be kept in bed in a dark room during several weeks.' Despite this hardship, he was so anxious about his work that his students were asked to visit him in his murky bedroom to report progress and receive new instructions.

Predictably, Perkin was a formidably diligent student, and found the preliminary coursework rather easy. He sat near a window overlooking Oxford Street's horsedrawn carriages, and spent some time sharing common interests with a man called Arthur Church seated opposite him. 'We were both given to painting and were amateur sketchers,' Church remembered. 'I was introduced to his home and we began painting a picture together. This must have been soon after the Royal Academy exhibition of 1854, when I had a picture hung.'

Church had created his own domestic laboratory by converting a small aviary at his home, so was keen to see Perkin's makeshift chemistry room on the top floor at King David Fort, where he worked in the evenings and at weekends. Perkin liked to take his work home with him, particularly when, after the completion of his basic syllabus in 1855, Hofmann had honoured him by making him his youngest assistant. 'The students working at research seemed to me to be superior beings,' Perkin observed.

* On another visit Hofmann found one of his students making good use of the gas fires to cook his meals. 'At lunch time he used to grill sausages in the empty, scoured dish of the sand-bath . . . or bake ham and eggs for him and his friends,' Hofmann's student Volhard recalled. 'Hofmann had often observed the not-quite chemical-smelling scent; one day he followed it and appeared quite unannounced in the makeshift kitchen . . . He dealt with his English pupil in masterly manner. No word of reproach, but he kept him busy until the last sausage was wholly charred.'

Perkin's earliest tasks concerned the formation of organic bases from hydrocarbons, but he was more interested in the results of his next assignment which led to one of his earliest published papers. At the beginning of February 1856 he submitted to the *Proceedings of the Royal Society* a brief report 'On some new Colouring Matters' he had found with Arthur Church. 'This new body presents some remarkable properties,' they wrote. That substance, which they named nitrosophenyline, was the result of a experiment with hydrogen and a distillation of benzol. It produced a bright crimson colour, it dissolved in alcohol with an orange-red tint, and it changed to a yellowish-brown when diluted with alkali. They concluded that it had 'a lustre somewhat similar to that of murexide', the rich purple originally made from guano.

Although August Hofmann was keen to see his students publish (and had in fact communicated the above findings to the journal himself), he believed the colourful discovery was of little value. In one sense he was right, for Perkin and Church could suggest no practical application for their new colour, and so they resumed other work. But it is significant that the pair, both painters, should be alert to what others might consider merely a pretty coincidence.

Hofmann faced other dilemmas. Many of the wealthier patrons of his college were concerned that chemistry was not producing results that would be beneficial to their well-being. And every landowner who had been excited by Justus Liebig's crusade was soon disappointed that the institution they supported was not, after all, their salvation. Subscriptions dwindled, and the college was forced to merge with the School of Mines; some students in Perkin's third year gained entry with the sole aim of improving coal extraction.

Even in 1856, there was much debate, and much disquiet, about the true virtues of pure chemistry. Triumphant practical men simply distrusted men of science. The success of the Great Exhibition of 1851, at which the magnificent Crystal Palace in Hyde Park hosted the most impressive display of booming mechanics that anyone had seen, suggested that progress would be forged merely by the continued application of cheap and copious steam power.

The problem with studying pure chemistry, on the other hand, was that the endeavour seldom produced anything remotely useful.

In the annual report of the Royal College published in 1849, August Hofmann revealed that one of his most cherished ambitions was to show how well the study of chemistry could produce the artificial synthesis of natural substances. He admitted that this involved an uncertain mix of supremely patient application and great good fortune. Indeed, Hofmann and his students were simply grasping for great things in the manner of skilled artists painting with untried materials. 'Perhaps we will be lucky,' Hofmann said.

By about 1830 it was becoming clear that all the substances isolated from plant and animal sources contained the elements carbon, hydrogen and oxygen, and often nitrogen and sulphur (the science of organic chemistry is essentially the chemistry of carbon compounds). A simple chemical compound was described by the combination of its elements. At school, Perkin would have learnt the basics: the elements were represented by chemical symbols such as C (carbon), H (hydrogen), O (oxygen) and S (sulphur), where an element is a substance that combines with others to form compounds, but which cannot be broken down into any simpler substance itself. When two or more elements combine, it is the atoms of the different elements that join together, forming molecules. Each molecule of a compound contains the same number of atoms as every other molecule of the compound. In the most rudimentary example, H_2O, the chemical symbol for water, thus contains two hydrogen atoms and one oxygen atom. It was not yet known that in some elements – such as the oxygen in the air – the atoms can join together to make molecules without other elements being involved.

One substance Hofmann wished to make in the laboratory was quinine. Quinine was the only treatment found to be effective against malaria, and in the middle of the nineteenth century malaria was a problem that determined the size and prosperity of an empire.

Malaria is an ancient disease, and perhaps the ruin of ancient civilisations. The fortunes of Rome and the Campagna have been tracked against its prevalence. It became widespread after the Second Punic War at about 200 BC, and declined during the days of the Roman Empire until the end of the fourth century AD. But it then reached epidemic proportions, and hampered colonisation until its decline shortly before the Renaissance.

The term malaria – the misleading literal translation of the Italian 'bad air' – was probably first used in English in the 1740s, when Horace Walpole described 'a horrid thing called the mal'aria that comes to Rome every summer and kills one'. Before this, its presence was defined by the catch-all diagnosis of fever or ague.

In Hofmann's day, malaria was the grave concern not just of Asia and Africa. France, Spain, Holland and Italy were still intensely malarious, although, as elsewhere in this period, it was not always possible to define precisely how many deaths were due to other fevers such as cholera. It was common in Russia and the Western Territory of Australia, and in America the disease was prevalent in the swamplands of the Carolinas, Florida and New Orleans. During the Civil War, malaria was the chief cause of death in the Southern States, and there were hundreds of thousands of cases in New York and Philadelphia – the situation only improving with the clearing and development of land.

In England, where, it was believed, malaria had been responsible for the deaths of James I and Cromwell, the disease was still rampant in the 1850s. The worst areas were the Cambridgeshire Fens and Essex marshes, and in *Great Expectations* Dickens depicted the marsh agues around Pip's home in Medway, Kent. Between

1850 and 1860, tens of thousands of people were admitted to St Thomas's Hospital diagnosed with ague (malarial fevers), and in 1853 it accounted for almost 50 per cent of all admissions.

British imperialists found malaria to be the greatest hindrance to colonisation. In Kilimane, for example, David Livingstone found the mosquitoes 'something dreadful', and described how he 'had an opportunity of observing the fever acting as a slow poison. '[Victims] felt out of sorts only, but gradually became pale, bloodless and emaciated, then weaker and weaker, till at last they sank more like oxen bitten by tsetse than any disease I ever saw.'

In India, where vast swathes of the country remained uncultivated because of malaria, the British Army were sending back reports of human devastation. It was suggested that the disease acted as a natural form of population control. Amongst adults, about 25 million struggled with the chronic nature of the disease, and about 2 million died annually. Horrified army officers reported grim symptoms. Patients suffered from chills, convulsions, burning temperatures, muscle pains, nausea, vomiting and delirium. Many died in a coma; many more found that their illness returned intermittently. The British naturally blamed the natives, despite the fact that their own policy of serfdom and the land policy of the East India Company compounded the severity of the problem.

Even in the 1850s, no one was sure what caused this disease. There were plenty of theories, most involving marshlands and airborne infection, but, despite vague hunches from Livingstone and others no one had yet made the scientific link with mosquitoes.

Treatment for malaria was more straightforward, although often difficult to secure. Up until 1820, when the French chemist–pharmacists Pelletier and Caventou isolated quinine from cinchona bark, many physicians still proffered such remedies as three days of blood-letting, or treatment with mercury, or three bottles of brandy. The superstitious believed that carrying a spider in a nutshell, or eating one, would cure the disease.

But this was the era of the new alkaloid. Cinchona bark (and roots and leaves) contained not only quinine (named after the Spanish spelling of 'kina', the Peruvian word for bark), but also cinchonine, and in the next two decades, two more alkaloids were isolated from the tree, quinidine and cinchonidine. Each had a slightly different molecular structure, and none was quite as effective against malaria as pure quinine (but nonetheless sold as such). In the same period, the two Frenchmen also isolated the strychnine from St Ignatius's beans, and other chemists found other alkaloids – caffeine in coffee beans and codeine in opium.

Quinine was in limited supply, and thus expensive. The cinchona tree is about the size of a plum tree with leaves like ivy, and was found almost exclusively in Bolivia and Peru. By 1852, the Indian Government was spending more than £7,000 annually on cinchona bark, and £25,000 for supplies of pure quinine. The East India Company was spending about £100,000 annually. Predictably, this was not intended to treat the poor, and still bought nothing like the 750 tons of bark required by the British army in India alone.

The clamour for quinine from the great European imperialists was immense, and Britain and Holland mounted costly attempts to grow cinchona seeds in India and Java; the British tried to grow the tree for commercial use in Kew Gardens. The initial planting missions failed, as explorers would often plant the wrong seeds in the wrong place. Some did get rich on the disease, the most notorious being John Sappington who marketed Dr Sappington's Pills in the Mississippi valley by persuading local churchmen to ring their bells in the evening to remind people to take them. Sappington had capitalised on the one fundamental property of quinine – its scarcity – and had added other worthless substances to his pills to make his supplies go further. In London and Paris, the cost of bark was about £1 per pound, but as it took approximately 2 lb of bark to treat each person, only the well-off got better. When, in the 1840s and 1850s, hundreds of thousands began demanding quinine as a prophylactic, it was clear that it had become the most desirable drug in the world.

In his room in Oxford Street, August Hofmann had a theory as to how quinine might be made in the laboratory. To his credit, he seems not to have been interested primarily in the fortune to be made by such a discovery. He had noted how naphtha, which he called the 'beautiful' hydrocarbon, produced in great quantities in the manufacture of coal-gas, may be converted by a relatively straightforward process into a crystalline alkaloid known as naphthalidine. This substance was found to contain 20 equivalents of carbon, nine of hydrogen and one of nitrogen.

Coal-gas contains more than 200 different chemical compounds, although only a few of them were known to Hofmann and his students in 1850. These are split between hydrocarbons (which include naphthalene, benzene, and toluene) and those compounds containing oxygen (the most important being phenol or carbolic acid).

Hofmann believed that, as the formula for quinine differed from that for naphthalidine by only two additional molecules of hydrogen and oxygen, it might be possible to make quinine from the existing compound just by adding water. 'We cannot, of course, expect to induce the water to enter merely by placing it in contact,' he wrote. 'But a happy experiment may attain this end by the discovery of an appropriate metamorphic process.'

William Perkin was only eleven when Hofmann published this theory, and he read it only after he was admitted to the Royal College in 1853. He soon recognised the importance of the idea. 'I was ambitious enough to want to work on this subject,' he recalled, and was motivated further three years later by Hofmann's chance remark that artificial quinine was now surely within their grasp. What he had not grasped was that the apparent simplicity of quinine's constituent parts would so thoroughly conceal the hidden complexity of their architecture. The 'happy experiment' desired by his mentor would not be forthcoming, or at least not in the way he had anticipated.

4
THE RECIPE

She said she was going to do it, and by golly, on Thursday, she did it. Because she is the first female secretary of state of Missouri, Judi Moriarty changed the color of the state manual to . . . mauve.

For those who don't know, mauve is a delicate shade of purple.

'I wanted a color that represents me and made a statement,' Moriarty said when introducing the new state manual. 'It's in good taste, and it has a lot of beauty.'

St Louis Post-Dispatch, 1994

In the first months of 1856, Gustave Flaubert began *Madame Bovary*, Karl Bechstein opened his piano factory, the plans for the bell Big Ben were drawn up at a foundry in Whitechapel and Queen Victoria instituted the Victoria Cross. During the Easter holidays of that year, August Hofmann returned briefly to Germany, and William Perkin retired to his laboratory on the top floor of his home in the East End of London. Perkin's domestic workplace contained a small table and a few shelves for bottles. He had constructed a furnace in the fireplace. There was no running water or gas supply, and the room was lit by old glass spirit lamps. It was an amateur's laboratory, an enthusiast's collection of stained beakers and test tubes and rudimentary chemicals. The room smelled of ammonia. The table on which he worked was stained with spillage from previous efforts, and probably of ink as well. He was surrounded by landscape paintings and early photographs, and by jugs and mugs and other domestic trinkets that were as alien to a laboratory as delicate soda crystals were to any other house in this smoky residential neighbourhood. It was an unexpected setting for one of chemistry's most romantic and significant moments.

Looking back, Perkin adopted a rather nonchalant tone to describe his actions. 'I was endeavouring to convert an artificial base into the natural alkaloid quinine, but my experiment, instead of yielding the colourless quinine, gave a reddish powder. With a desire to understand this particular result, a different base of more simple construction was selected, viz. aniline, and in this case obtained a perfectly black product. This was purified and dried, and when digested with spirits of wine gave the mauve dye.'

In effect, the discovery at that time of one apparently simple molecule could rarely claim such a far-reaching impact on the development of science and industry. The room in his father's house afforded views of the ships in the London docks, and of the London and Blackwall Railway, an inspiring vision of travel and progress. But Perkin's view of the distance held no glimpse of the future, no vision of the Lancashire factories 200 miles away which soon would reverberate with the sound of his invention.

The chemistry was simple, involving the then popular 'additive and subtractive' method: find a compound that looks similar to the one you are trying to create – in this case, Perkin chose allyltoluidine – and used two standard processes, distillation and oxidisation, to alter its formula by adding oxygen and removing hydrogen (in the form of water). It was a naive manoeuvre.

Most chemists, particularly those trained by Hofmann at the Royal College, would have thrown the reddish powder into a rubbish bin, and begun again. It was Perkin's intuitive talent – an enquiring mind in an unsupervised laboratory – that led him to experiment further, and test the effect of this procedure on aniline. And it was a mark of his skill that, in analysing the crude black product that resulted, he was able to separate out the five per cent that contained his colour.

By the time Perkin found mauve, aniline had been linked with colorants and colour-producing reactions for thirty years. The liquid had first been discovered by the Prussian chemist Otto Unverdorben in 1826, one of several products isolated from the distillation of natural vegetable indigo. Some years later the chemist Friedlieb Runge obtained it from the distillation of coal-tar, and found it gave a blue colour when combined with chloride of lime. But such colours were considered to have no practical use. In the unlikely event that a scientist would have thought a particular tint might be useful in the dyeing of a woman's dress, they would most certainly have believed such fripperies unworthy of their calling.

But Perkin was excited about his unexpected find. Chemists blundered every day; partly, that was the nature of their job. But only occasionally did their errors lead them in interesting directions. Perkin stained a silk cloth with his discovery, and did little more than admire the new shade. It was, he realised, a brilliant and lustrous colour, and he found that it did not fade with washing or prolonged exposure to light. The problem he faced was what to do with it next. 'After showing this colouring matter to several friends, I was advised to consider the possibility of manufacturing it upon the large scale.'

One of these friends was Arthur Church, with whom Perkin discussed the seemingly insurmountable problem of making more than a small beaker of his colour. Liquid aniline was hard to obtain in quantity, and expensive; Perkin had never set foot inside a factory, and knew nothing of manufacturing chemicals outside the laboratory; and he knew no one in the textile or dyeing trades to whom he could turn for advice.

Both Perkin and Church knew that their mentor would disapprove of any schemes not directly connected with research. They resolved not to tell Hofmann about mauve when he returned from

Germany, certainly not until Perkin had established its exact properties and had conducted further experiments.

For this, Perkin moved to slightly largely premises – a hut in his garden. He enlisted the help of his brother Thomas, and together they made several batches of mauve, each purer and more concentrated than the last. Through a friend of his brother, Perkin learnt the name of a highly regarded dye works in Scotland, and decided to send the owner some samples of cloth. He received a lengthy reply from a man called Robert Pullar in the middle of June, and his tone was encouraging.

'If your discovery does not make the goods too expensive it is decidedly one of the most valuable that has come out for a very long time. This colour is one which has been very much wanted in all classes of goods and could not be had fast on silk and only at great expense on cotton yarns. I inclose you patterns of the best lilac we have on cotton. It is done by only one house in the United Kingdom, Andrews of Manchester, and they get any price they wish for it, but even it is not quite fast, it does not stand the tests that yours does and fades by exposure to air.'

Pullar was twenty-eight, and was later described by a general manager of his company as possessing 'a mind always looking forward for something new and better'. His large dye works in Mill Street, Perth, had recently received a royal warrant, and now advertised itself proudly as silk dyemakers to the Queen. Robert Pullar liked to quote Faraday: 'Without experiment I am nothing; still try, for who knows what is possible.' Perkin had been lucky in his choice of adviser; he was to discover later that not all dyers or printers were as progressive or encouraging.

Pullar explained to Perkin that he could not put a price on the colour, not until he had tested it himself in a dyeing vat. 'If the quantity of yarn or cloth that could be soaked in one gallon of your liquor would take up all the colouring matter in that gallon, then I would say that the price would be much too great . . .' If this happened, the dyestuff required to colour one pound of silk or cotton would cost about five shillings – 'far too much for a manufacturer to pay'.

Pullar offered to help Perkin in any way he could, and regretted that he did not live nearer London to meet him in person. 'We are always very desirous here to have every thing new, as we do a large trade in manufacturing and a new colour in the goods is of great importance.'

Perkin showed this letter to Arthur Church, who encouraged him to take out a patent immediately. But there was a problem with Perkin's age, as patents were usually only granted to those over twenty-one. He sought counsel's opinion, and was advised that since a patent was a gift from the Crown, the matter of age should be immaterial. Perkin filed his application at the end of August 1856, when he was eighteen. But then he began to wonder: what good would it do him? Just how much was a new colour worth?

New colours had been discovered by chance since ancient times, and some magnificent myths had evolved. A sheep dog belonging to Hercules, while walking along a beach in Tyre, bit into a mollusc which turned his mouth the colour of coagulated blood. This became known as Royal or Tyrian purple. It brought prosperity to Tyre around 1500 BC, and for centuries remained the most exclusive animal dye money could buy. It was the colour of high achievement and ostentatious wealth, and came to symbolise sovereignty and the highest offices of the legal system. Within Jewish practice, the dye was used on the fringes of prayer shawls; in the army, the wearing of purple woollen strips was used to denote rank. Purple was also the colour of Cleopatra's barge, and Julius Caesar decreed that the colour could be worn only by the emperor and his household.

It was prohibitively expensive. The molluscs – *Murex brandaris* from the Italian coast or *Murex trunculus*, located first on the Phoenician coast – were drained of their glandular mucus in their thousands to make a single robe. Pliny described how, during autumn and winter, the shellfish were crushed, salted for three days and then boiled for ten. The resultant colour resembled 'the sea, the air and a clear sky', suggesting that Tyrian purple defined not one particular shade but a rich spectrum from blue to black. The dying process varied from port to port, and might have water or honey mixed in to achieve different hues.

Of the other animal dyes the most popular was cochineal, the crimson dye from cactus insects. Introduced into Europe by the Spanish from Mexico (then New Spain) in the sixteenth century, it was widely used as cloth dye, artists' pigment, and much later a food colorant, but again required a huge seasonal harvest – about 17,000 dried insects for a single ounce of dye. What may have been the first English dye house was established for cochineal in Bow, east London, in 1643, and the scarlet became known as Bow-Dye and was described in terms of bruised flesh.

Vegetable dyes tended to be cheaper, and in greater supply. In Perkin's day the most common were madder and indigo, the ancient red and blue dyes used for cloth and cosmetics. Madder, from the roots of some 35 species of plant found in Europe and Asia, has been found in the cloth of mummies and is mentioned by Herodotus, and is probably the first dye to be used as camouflage – Alexander the Great spattering his army with red to persuade the Persians that they had been critically wounded in earlier battle. In 'The Former Age', *c.*1374, Chaucer depicts the idea of man's early innocence when

> No mader, welde, or wood [woad] no litestere [dyer]
> Ne Knew; the flees [fleece] was of his former hewe.

Indigo, used not only as dye and pigment but also an astringent lotion, derived from the leaf of Indigofera tinctoria, a shrub-like plant that was soaked in water and then beaten with bamboo to hasten oxidation. During this process the liquid changes colour from dark green to blue, when it is then heated, filtered and formed into a paste. Before the colonisation of America, it came predominantly from India in the form of dye-cakes, and this ancient derivation held firm to the time when Perkin could observe the colour in women's fashions in the West End.

There were several other important plant dyes – carthamus, woad, saffron, brazilwood and turmeric – but even these represented an extremely narrow range of colours, confined variously to red, blue, yellow, brown and black. Woad, again known to Pliny and used commonly by ancient Britons as a facial and body dye, contained a similar colouring matter to indigo, although derived from a different plant and containing about one-tenth the tinctorial power.

Throughout much of the eighteenth century the greatest advances in dyeing technique were made in France, but between 1794 and 1818 an American working in London called Edward Bancroft claimed many significant improvements. Bancroft patented three new natural dyes, including the yellow quercitron, and wrote the first scientific treatise on dyeing in English. His *Experimental Researches Concerning the Philosophy of Permanent Colours* combined exact chemical observations with personal anecdotes: he noted, for example, how his favourite purple coat hardly faded though he wore it for several weeks. Bancroft had a further claim on posterity, as he was later exposed as a double agent during the American Revolution, working both for the British government and for Benjamin Franklin.

The process of dyeing cloth had not changed much in centuries, and the most skilled practitioners had passed complex and guarded procedures through generations. But in New York in 1823, William Partridge published *A Practical Treatise on Dyeing of Woollen, Cotton and Skein Silk, with the Manufacture of Broadcloths and Cassimeres Including the Most Improved Methods in the West of England*, for thirty years the standard text, in which all the most popular dyes were disclosed like magicians' secrets and presented like cookery recipes. To prepare the fastest blue, for example, you would need an English vat containing 'five times one hundred and twelve pounds of the best woad, five pounds of umbro madder, one peck of cornell and bran, the refuse of wheat, four pounds of copperas, and a quarter of a peck of dry slacked lime'.

There were detailed descriptions of how to prepare the lime, followed by directions to chop the woad into small lumps with a spade, and gradually add other ingredients to water set at 195 degrees Fahrenheit. The instructions ran on for several pages. 'The vat should be set about four or five o'clock in the afternoon, and be attended and stirred again at nine o'clock the same evening,' before being cooled. By this stage the result should be bottle-green. The dyer was then directed up again at five in the morning, and told to add more lime or indigo to lighten the colour. Bubbles and skin and increasing thickness would denote a good fermentation, which should then be boiled again and cooled, and boiled and cooled, and more lime added, and then it was time for the wool dipping. This was where matters became complicated. You really needed two vats of woad, one two months old, the other new, and the wool should be dipped in each in turn. The temperatures of the dye should be finely held at 125°F–130°F, then cooled overnight, then heated to 155°F–165°F, and then more woad added, with more lime, bran, madder and indigo. If the vats were skilfully managed it should colour 220 pounds of wool every week; within six weeks, the dyer should have four hundred pounds of dark blue wool, two hundred of half-blue, and two of very light. But this was only attainable if the very best woad and indigo were used, and here there were problems: 'There is probably no article more uncertain in its strength and quality than woad,' Partridge concluded. He advised buying only the very strongest, as 'any considerable variation in this particular will prove very disastrous to the operator, however skilful he may be in his profession, and will be altogether ruinous to a young beginner'.

As with cinchona bark, the supply of plant dyes was often limited to specific regions and hampered by a nation's attempts to monopolise production. Clothes manufacturers were forced to use the colours available in the dyers' vats; trends in colour were fashioned less by taste than by the vagaries of war and efficiencies of foreign ports. It stood to reason that a colour you could make on demand in a laboratory somewhere, with a constant strength and purity, would surely be worth an awful lot of money.

Initially, Perkin called his discovery Tyrian purple, the better to elevate its worth. His detractors, those who believed his discovery insignificant, preferred to call it purple sludge. Chief amongst these was August Hofmann, who learnt of Perkin's new colour after the summer holidays, along with some distressing news of his protégé's future. The two arranged a meeting, during which Perkin told Hofmann that he was considering manufacturing mauve commercially. He also said that this would require him to leave the Royal College of Chemistry. 'At this he appeared much annoyed,' Perkin recalled at a memorial meeting to mark Hofmann's death in 1892. '[He] spoke in a very discouraging manner, making me feel that perhaps I might be taking a false step which might ruin my future prospects.'

The objection caused a serious rift between them – probably the first cross words they had exchanged. 'Hofmann perhaps anticipated that the undertaking would be a failure, and was very sorry to think that I should be so foolish as to leave my scientific work for such an object, especially as I was then but a lad of eighteen years of age. I must confess that one of my great fears on entering into technical work was that it might prevent my continuing research . . .' *

* While Hofmann objected to Perkin's new obsession, it was not solely due to his pursuit of a practical application of his learning. Hofmann himself was involved in several such projects: in 1854 he analysed the spa waters of Harrogate for the Harrogate Water Committee; he sat on the chemical sub-committee assigned to examine the decay of the limestone and dolomite structure of the Houses of Parliament (no solution agreed upon); and in 1859 the Metropolitan Board of Works asked Hofmann to consider the possibilities for deodorisation.

Hofmann and his colleagues would have found it hard to imagine how one of the most promising scientific careers could be summarily abandoned in pursuit of a colour. Chemists came across new colours at random almost every week, and just as easily dismissed them as being an undesirable or irrelevant side-effect of their missions. Besides, some chemists had deliberately produced artificial dyes before mauve, and had observed how well they had coloured silk or wool, but had not attempted to manufacture them in commercial quantities. The first had been the picric acid made by Woulfe in 1771 from indigo and nitric acid (it dyed silk bright yellow), and in 1834 Runge had used carbolic acid to make aurin (a red colour), and pittacal (a deep blue) was obtained from beechwood tar. Other colours encouraged the development of implausible histories, not least murexide, which surfaced in small quantities in Manchester dye works in the 1850s and was said to come from the excrement of serpents (rather than its true source, bird-droppings). But the quantities of synthetic dyes in use at the time of the Great Exhibition of 1851 was so small as to not merit any mention in the huge accompanying Reports.

Then there was the bright crimson produced by Perkin and Church some months before, again considered unworthy of further exploitation. Perkin's purple might have been cast aside in a similar manner were it not for the further encouragement he received from Robert Pullar in Perth towards the end of 1856.

The scale of Pullar's dye works must have seemed an impressive place to a young man unfamiliar with industrial practices. The presence of scientists, however, was nothing new to print works, and some had employed their own textile chemists from 1815. In fact, Perkin's discovery came at a time when the state of technical advance in Britain's dye and printing works was ideally poised to exploit it. Production levels in the textile industry were increasing at unprecedented rates. Exports in the calico business, for example, increased fourfold between 1851 and 1857, from about 6,500,000 items to 27 million. Employment in the silk industry doubled to 150,000 people between 1846 and 1857. At one of the many jubilee celebrations of Perkin's discovery, the chemist C. J. T. Cronshaw told a gathering of the Society of Chemical Industry: 'If a fairy godmother had given Perkin the chance of choosing the precise moment for his discovery, he could not have selected a more appropriate or more auspicious time.'

This was not only true of the position of Britain's dye works. Perkin could only have discovered mauve when he did because of the particular state of chemical knowledge. He was born not long after the Cumbrian chemist John Dalton had theorised that atoms combine with each other in definite numbers, thus leading to the establishment of chemical formulae. But Perkin conducted his early experiments at a time when so much was yet unknown, thus allowing for his productive error over the synthesis of quinine. If Perkin had been born twenty years later, he would have known how fruitless his search would have been, and thus would not have blundered into mauve. John Dalton, incidentally, died twelve years before Perkin's discovery, but the beauty would have been lost on him anyway: in 1794 he had been the first person to describe colour blindness – his own.

The principal reason that August Hofmann would have failed to share Perkin's enthusiasm for his new colour was because he would not have been unduly surprised by it. Even before he came to London he had heard Liebig predict that artificial dyes would someday be made from a substance such as aniline. But the roots of his disapproval lay in the current relationship between pure and applied science, which really meant the relationship between science and industry, two worlds set against each other by deficiencies in education.

In 1853, Lord Lyon Playfair had travelled through Germany and France at the request of the Prince Consort, specifically to report back on the state of foreign scientific and technical education. His analysis was damning: the great universities of Europe had already forged a strong connection between laboratory work and industry, whereas in industrial Britain he found only an 'overweening respect for practice and contempt for science'. He found the greatest culprit to be the severe shortcomings in basic teaching. Playfair feared the impact on Britain in the event of free trade, suggesting that when 'the raw materials confined to one country become readily available to all at a slight difference in cost, then the competition in industry must become a competition in intellect'.

The Great Exhibition of 1851 inspired many lectures sponsored by the Society of Arts, and some of them singled out a peculiar irony: while Britain shook the world with its industrial clout, it was virtually alone in Europe in lacking a well-defined system of technical education.

The same year saw the opening of Owens College in Manchester, and at its inaugural gala the college's professor of chemistry Edward Frankland suggested that Britain's textile industry was ill-prepared for the future. Its pre-eminence in manufacture would only be maintained by far stronger links with men of science. 'The advantages of chemistry to the chemical manufacturer, the dyer and calico printer are almost too obvious to require comment,' he said. 'These processes cannot be carried out without some knowledge of our science, yet with the exception of some few firms . . . this knowledge is too often only superficial, sufficient to prevent egregious blunders and ruinous losses, but inadequate to seize upon and turn to advantage the numerous hints which are almost sure to be constantly furnished in all manufacturing processes.'

The Chemical Society, founded in 1841, drew its few hundred members from manufacturing and academic backgrounds, and prided itself on the links between the two. In 1853, the president of the society, Frank Daubeny, seemed to express relief when he informed his members that Professor Robert Bunsen's work on volcanic eruptions could be used as 'undeniable evidence of the extensive utility of our pursuits'. Four years later, the new president W. A. Miller spoke of the invention of mauve as further proof of the burgeoning usefulness of their skills. 'One of our Fellows, Mr Perkin, has afforded me the opportunity of bringing before you the results of a successful application of abstract science to an immediate practical purpose.' At the time, he could hardly have known of the immense implications of this observation.

In fact, the successful application was still some months away, but when it came it did little to placate those who believed that Perkin's intellect could be better employed elsewhere. Even in 1862, it appeared that Hofmann accepted Perkin's breakthrough only very grudgingly. After visiting the International Exhibition that year in London, he still wished that 'the care and time involved in an undertaking of such magnitude may not divert [Perkin] from the path of scientific enquiry, for which he has proved himself eminently qualified'. Such a pure attitude ran counter to the dominant industrial ambition of the age: the pursuit of wealth.

In retrospect, it appears that Perkin shared some doubts about his commercial ambitions, though not for fear of being thought greedy. He resolved to regard his foray into industry as a means to an end. Writing to his friend Heinrich Caro, he stated that at the time of his discovery, 'for a scientific man to be connected with manufacturing was looked upon as infra dig.' Scientists who crossed the line were treated as pariahs, betrayers of their calling. Perkin was worried that, should he fail, there would be no way back. 'Even poor Mansfield, as soon as he started to be a manufacturer, sold his scientific instruments (I have his balance which I purchased from him) evidently with the idea that his research days were over,' Perkin wrote. 'This public opinion and example made me dread becoming a manufacturer, because research was the principal ambition of my life, and I determined so far as in me lay that I would not give this up, whatever I did.'

At the time, however, he kept this desire very much to himself, and was treated by some with disdain. 'It was said that by my example I had done harm to science and diverted the minds of young men from pure to applied science, and it is possible that for a short time some were attracted to the study of chemistry from other than truly scientific motives.' In other words, Perkin's discovery affected the whole nature of scientific investigation: for the first time, people realised that the study of chemistry could make them rich.

How to Make Mauve: a modern method

Caution: Petroleum ether is extremely flammable. All evaporations should be performed under hoods (fume cupboards). No naked flames in the lab, please. Disposal of all chemical wastes should follow the standard procedures.

You will need:

2.3ml of water in a 5ml conical flask, to which add
52μl of aniline
60μl of o-toluidine
122mg of p-toluidine
600μl of 2N sulphuric acid

Stir, using a large spin vane, until the reactants have dissolved, heating gently if necessary. After solution, add 30mg of potassium dichromate in 160μl of water.

Stir for two hours. Very soon after the addition of $K_2Cr_2O_7$, the solution will turn a vibrant purple. At the end of the reaction time, use a Pasteur filter pipette to draw off the liquid portion, which can be discarded.

Transfer the solid to a ceramic filter with a seated filter paper already in place. Using gentle suction filtration, wash the dark solid with distilled water until the washing is clear. Dry the remaining solid in an oven at 110°c for 30 minutes. Then wash the solid with petroleum ether until the washings are clear. Dry again for 10 minutes at 110°c.

Wash the remaining solid with a 25 per cent methanol/water solution until the liquid runs clear, being very careful not to contaminate the product. Evaporate this aqueous/alcoholic solution, transferring to a 5ml conical flask as soon as the total volume allows. After evaporation is complete, add 300µl of 100 per cent methanol to the remaining solid, shake to dissolve any soluble materials, and use a filter pipette to transfer the liquid to a clean 3ml flask. Carefully evaporate the liquid in a conical flask until it has a volume of 30µl or less. As the solution volume gets smaller, the purple colour should grow more intense. This final methanol solution contains the ultimate product – a 2mg yield of mauve.

This isn't much mauve. To get more, you might like to ask all your friends or an eager class of chemistry students to conduct the same experiment, and then pool all the methanol solutions, before evaporating this *en masse*. A small piece of cotton cloth can then be dipped in the solution, then rinsed in water and dried. It makes a nice pocket handkerchief; the solution will colour three slim bow ties.

How to make a Nesselrode: a traditional method.

The Nesselrode pudding, a frozen tower of chestnut, fruit, custard and cream, is the creation of Monsieur Mouy, chef to Count Karl Vasilyevich Nesselrode, the nineteenth-century Russian statesman best known for his role in the creation of the Holy Alliance. The pudding appears, to some acclaim, in one of the lengthy dinner parties in Proust's *A la recherche du temps perdu*, but these days is seldom made, largely due to health considerations.

The Nesselrode has been described as 'a two-day event, not counting the evening with the orange peel'. One day takes place in the freezer.

You will need (for six people):
½ cup chestnut purée
¼ cup glacé cherries
½ cup candied orange peel
½ cup Marsala
½ cup each currants and sultanas
1 big splash maraschino
2 cups whipping cream
Custard: 2 cups milk, 5 eggs, ¾ cup sugar

Chop the candied fruit and work in the Marsala. Soak the currants and sultanas in warm water and dry. Heat the milk to boiling. Separate the eggs, discard the whites but place the yolks in a bowl and add the sugar, beating vigorously until the mixture is frothy. Pour the milk over the mixture in the bowl, then place the contents of the bowl in the saucepan. Stir frequently and cook over low heat until the custard thickens and then strain through a sieve. Mix the chestnut purée, the maraschino and the custard together well, then add all the fruit. Whip the cream until stiff, and fold into the mixture. Pour into a mould lined with waxed paper or greased up like an oily whippet. Cover with foil and freeze for a full day.

It is a costly dish, but it will feed a big celebration. For a special effect, such as a birthday or where lettering is required to inform people why they have come, the surface may be coloured with a myriad of dyes. Try Annatto, Quinoline Yellow, Fast Green, Citrus Red No. 2, E160c, E100, E150d and E129.

5
HINDRANCE AND SYNTHESIS

Patrick mixed paints – a delicate shadowy mauve, a scarlet, a rich blue, a pale sharp green. The paintings, when they arrived, were done suddenly and fast. I watched, from inside my head. Patrick would always smile apologetically, and both of us would laugh nervously, and then his face would set into a detached, slightly furious look, and he would take a stab at the canvas, and then a rush.

A square head appeared, and a decorative trellis of flowers. Various faces, shadowed in the delicate mauve, existed for a moment, and then were wiped away. I was fascinated by how the ghosts of the expunged forms continued to exist and to make the subsequent versions more complex and substantial. Purple is Patrick's favourite colour. It is not mine. But I became entranced by the shadowy half-depths of that particular mauve running across the canvas.

A. S. Byatt on being captured by Patrick Heron, *Modern Painters*, 1998

The way Robert Pullar remembers it, William Perkin arrived in Perth, Scotland, in the early winter of 1856 with some problems on his mind. He explained to Pullar that he had severe doubts about his decision to leave the steady world of research for one of commercial industry. He had misgivings about his ability to compete in business, and this insecurity led to more fundamental concerns. Just how valuable would mauve be? How would dyers react to a complete change in the way colour was derived? Would anyone be willing to adapt to new methods of dyeing after employing the same techniques for generations? And although mauve successfully dyed silk, why was it proving so much harder to apply the new colour to cotton?

Writing many years later in *The Dyer*, Laurence Morris listed six hindrances to Perkin's future success, a daunting tally for an eighteen-year-old. (1) There was no reason to believe that capital could be found to launch the venture. (2) There was no guarantee that dyers and printers would use the colours at all, let alone in the quantities required to justify the building of a factory to manufacture it. (3) The raw materials required to make it were largely crude chemicals in limited supply. (4) No site was available to build a factory. (5) Methods of applying the new colour to most textiles had yet to be worked out. And (6) Perkin did not have any experience of what he was about to do. In all ways, Perkin appeared to be quite out of his depth. Making a colour was one thing; bringing it to the world quite another.

To reassure each other, Perkin and Robert Pullar spent a little over two weeks conducting experiments and touring Scottish dye works. The results were disappointing. Mauve would not dye cotton without a mordant (an intermediate fixative that allowed the colour to 'bite' into the material). Unlike the existing vegetable dyes, mauve was not acidic, and all the existing mordants Pullar possessed failed to work on a synthetic dye; on cotton, mauve faded dramatically after each wash. The conservative printers and dye technicians they met in Glasgow were consistently brusque and dismissive. 'Ach, ye're wrang this time, Mr Robert,' one printer remarked on their travels. 'D'ye think anything will take the place o' madder?'

'It was a very discouraging day,' Perkin noted as his pessimism deepened. 'Although the colours were admired, that terrible IF respecting the cost was always brought forward.' He estimated that he would be unable to supply purified solid mauve for less than £3 per ounce – this at a time when an ounce of platinum cost just over

£1. His problem was that to make one ounce of mauve he required about 400 pounds of coal.

To his advantage, mauve appeared to be the most intense colour the dyers and printers of Scotland had seen. In one experiment, Perkin demonstrated that one part of mauve dyed 630,000 parts of water. 'Yet the printers who tried it did not show any great enthusiasm,' Perkin concluded, fearing that he was beginning to lose the support of his host. 'Even Messrs Pullar began to fluctuate in their opinion as to the advisability of erecting plant for its manufacture.'

Perkin returned to London to receive a curious note from August Hofmann. Resigned to losing him as a student, Hofmann realised that he might now be useful as a supplier of materials for his own researches. 'My dear Mr Perkin,' he began. 'Since you manufacture aniline now on rather a large scale, would you be so good as to let me have a quart or two of it. Of course I am most anxious to pay for it. I am very prepared for it and would thank you much for a speedy reply to this note.' It is not known how Perkin responded. But despite Hofmann's assumption, Perkin was having trouble securing enough aniline for his own samples.

Shortly after Christmas, Perkin received another letter, this time with positive news. Robert Pullar had found one dyer who might be supportive. 'As I was unable to help [the application of mauve] forward myself,' Pullar later recalled, 'I introduced Sir William to a friend, Mr Keith, in Bethnal Green, the largest silk dyer in London at that time.'

Thomas Keith, who had made a big splash with a beautiful display of silks at the Great Exhibition of 1851, offered Perkin the two things the Scottish dyers had not: faith and daring. 'He believed in it from the first,' Pullar said. After Keith had tried some of his own experiments with mauve, Pullar observed that these new shades 'are certainly the best I ever saw'. Shortly afterwards, both Pullar and Perkin independently found a tannin mordant with which to apply mauve to cotton and fasten it against water and light. Pullar wrote again to Perkin in triumphant tones: 'I suppose you are now fully decided to go on with it, from what I have seen there appears to be no doubt of its success.' But the optimism was premature.

'Time would fail me to enter into all the difficulties that beset the establishment of this unique industry,' Perkin wrote some years after the event. 'In fact, it was all pioneering work.' The first problem was financial: no one seemed willing to support his venture. Though Perkin had no experience of the banking market, his older friends were also unable to secure capital for what investors regarded as a harebrained and 'unnecessary' scheme. But after several months of rejection, Perkin's father decided to take on the venture himself. 'Although he had been disappointed at my becoming a chemist,' his son remembered, '[he] nevertheless had so much confidence in my judgement that he very nobly risked most of the capital he had accumulated by a life of great industry in order to build and start works for the production . . .' Perkin's older brother Thomas Dix Perkin also abandoned his architecture course in order to help establish the dye works, and together the three of them – now registered as Perkin & Sons – searched for a suitable site. Here too there were problems, and their original plans for land in the East End were thwarted by their ambition. They were seeking a large open area with good water supplies and transport connections, and potential for steady expansion. They faced objections from local planning authorities suspicious of the hazards of new chemical works; the newspapers were full of stories of noxious gases and poisoned streams. They had about £6,000 to spend, and found themselves pushed northwards, further and further from their home.

Their frustration was noted in a letter to William Perkin from John Pullar, Robert's brother, in May 1857. John Pullar was also in the dyeing business, working at Bridge of Allan, one of the factories Perkin visited on his trip to Scotland. 'I observe you have not yet begun to set the Thames on fire,' Pullar wrote. There was a certain poignancy in this phrase, as Pullars Dye Works had recently been badly damaged by a blaze. ('Had the wind been high,' Pullar noted, 'probably our whole works and machinery erected at such a cost of trouble and expense would have been destroyed.') Pullar was sorry to learn that Perkin had as yet been baffled in his attempts to secure a piece of ground for his works. 'I sincerely hope you may e'er long meet with a suitable place, and on reasonable terms – but I fear you will have to seek for it as you indicate near Manchester or Glasgow, or some such place where a chemical works is not such a bugbear as it appears to be to the benighted inhabitants in the vicinity of the Metropolis.'

Writing previously to Pullar, Perkin had told him that Thomas Keith had asked female friends what they thought about mauve, and their reaction had been effusive. Pullar perhaps misunderstood the size of the sample. 'I am glad to hear that a rage for your colour has set in among that all-powerful class of the community – the ladies. If they once take a mania for it and you can supply the demand, your fame and fortune are secure.'

Only a week after this letter arrived, Perkin & Sons at last secured a suitable site, but it was in an area unfamiliar to them – Greenford Green near Harrow, north-west of London in the country of Middlesex. The site was a meadow close to the Grand Junction Canal. At a little over six acres, it offered plenty of room for development. But it was a modest location from which to plan a revolution in colour.

The Perkins had found the spot through an advertisement placed by a woman called Hannah Harris, who had inherited the land from her husband Ambrose, the owner of the Black Horse public house at the edge of the site. Mrs Harris had only one condition about the sale of the land: no other pub should be built on it.

William Perkin's father began construction at the end of June 1857, a project that would last six months. The Perkins calculated that the production of mauve would require seven small and low buildings, three for direct manufacture and the others to house materials, an office, staff and a laboratory. During construction, the Perkins moved from Shadwell to a temporary home a short walk from the site, where William set up his chemical equipment in the small back wash-house and conducted new experiments on the artificial synthesis of various acids. He described these domestic labours as 'very difficult and painful', and poor ventilation meant he frequently abandoned his work when overcome with vapours.

The meadow often became waterlogged when the nearby brook burst its banks, but otherwise water served them well; there was a plentiful supply for manufacture, and the nearby canal carried barges that would, it was hoped, soon carry cakes of mauve dye to the fashionable printers in London and then to the continent.

During the months of construction, William Perkin travelled the country in search of suitable materials. As no one had worked with aniline on a large scale before, he reasoned he would have to make it himself. Originally he had hoped that he would be able to obtain sufficient quantities by the method which led to his original discovery, but concluded that this would make it too expensive (such a tiny amount was produced from so many tons) and was anyway too complex a procedure to conduct on a grand scale. He concluded he would make aniline from coal-tar-dreived nitrobenzene, although this too was problematic.

'Benzene at this time was only made to a very limited extent, as there was but little use for it,' he noted. After some searching, he ordered his main supplies from a chemical works in Glasgow at five shillings a gallon. 'In commencing this manufacture, it was absolutely necessary to proceed tentatively,' he observed, 'as most of the operations required new kinds of apparatus.' Thus he and his brother were obliged to design their own machinery, and the early models were primitive and highly combustible. One of Perkin's illustrations shows a cast-iron cylinder holding up to 40 gallons with a stirring tool at one end and a lid fastened by a cross-bar at the other. There were two funnels built into the top, one to let in benzene and sulphuric acid, the other to emit nitrous fumes. Significantly, the design of Perkin's equipment remained a prominent feature in Europe's dye works for eighty years. 'The nitration of benzene is not, of course, a process free from all danger,' Laurence Morris noted subsequently, 'and in those early, groping stages of manufacture it is a miracle that Perkin did not blow himself and Greenford Green to pieces.'

In less than six months after the building of the works had commenced, aniline purple was dyeing silk in Thomas Keith's dye house.

In 1858, Perkin combined a trip to Leeds with a visit to the annual meeting of the British Association. During his opening address, Richard Owen, the president, sensed a big change in the world of the chemist. Modern chemistry was on the cusp of unfathomable advances. 'To the power which mankind may ultimately exercise through the light of synthesis, who may presume to set limits?' he asked. 'Already, natural processes can be more economically replaced by artificial ones in the formation of a few organic compounds . . . It is impossible to foresee the extent to which chemistry may ultimately, in the production of things needful, supersede the present vital energies of nature.'

A few hours later, William Perkin gave a speech entitled 'On the Purple Dye Obtained from Coal Tar', a lecture he illustrated with samples of his work. He held up a sheet of silk, and then a skein of wool, and ran through the process of how they came to be. The assembled scientists declared themselves entranced by the manufacture and enamoured of the product.

One year later, Perkin was addressing the Society of the Arts on New Year's Day. He said of mauve: 'I will now tell you how it is made.' The process was clearly complex, and may have seemed so even when broken down into simple terms for his distinguished audience. The procedure took two days, Perkin said. The process combined aniline, sulphuric acid and bichromate of potassium, resulting in a fine black solution that was filtered to a soot-black powder. This contained various products apart from mauve, the most troublesome being a brown resinous substance that had to be removed with naphtha and methylated spirits. Perkin again had to design special apparatus for this process, and he had particular problems with the sealants and joints: 'It is remarkable the amount of difficulty and annoyance they caused.'

The substance he finally produced was placed in a still, where the spirit was distilled off. The remaining fluid was then filtered, washed with caustic soda and water and drained on another filter, leaving a very dark mauve paste. Perkin then presented his audience with a list of substances on a chalk board, an illustration of how much coal was required to produce so little mauve. The list began with 100 pounds of coal, from which was derived 10 pounds 10 ounces of coal-tar, 8½ ounces of coal-tar naphtha, 2¼ ounces of aniline, and only ¼ ounce of mauve.

Perkin claimed that one pound of his mauve could dye 200 pounds of cotton. He then produced a large bottle known as a carboy. He said there were nine gallons of water in it. He dropped in one grain of mauve, and illuminated the spectacular result with a magnesium lamp: the whole container turned mauve within four seconds. It was understood, for measuring purposes, that one gallon of water contained 70,000 grains, and that his bottle contained 630,000 grains. 'This solution,' he concluded, 'contains only one part of mauve to 630,000 of water!' He sat down to huge applause.

Perkin was used to this reaction, for he had conducted this demonstration numerous times on his travels. What Perkin did not tell them was the initial disdain that had greeted these samples when he had shown them to printers. 'The calico printers especially were not at all excited about it,' he wrote later. Their response sent him into a 'mild despair', and he was shaken by the printers' power to so quickly dismiss a new opportunity; he noted that their reaction was little changed from that of the Scottish dyers he had visited two years earlier, before he had abandoned the Royal College, before his father had sunk his savings into his factory.

At the same time, he faced another piece of bad news. He learnt that he had failed to secure the French patent for mauve because he had not registered within six months of his British patent. He soon found that at least one established dye works in Lyons had apparently copied his process and was producing a very recognisable colour.

For a few weeks at the beginning of 1858, the Perkin family experienced deep gloom. William Perkin must have wondered whether August Hofmann had been right all along: was he indeed throwing away the prospect of a brilliant career? Despite these doubts, he maintained eighteen-hour days, developing the factory, limiting explosions, improving his methods, trying to interest British printers in a simple invention. And then two things happened to change his life. Queen Victoria wore mauve to her daughter's wedding; and Empress Eugénie, the single most influential woman in the world of fashion, decided that mauve was a colour that matched her eyes.

6
MAUVE MEASLES

Knights of old broke each other's ribs, and let out each other's blood, dying happily among a heap of shivered armour, so that their ladies' colours still waved from their helmet, or sopped up the blood oozing from their gaping heart wounds; but you, Mr Perkins [*sic*], luckier than they, rib unbroken, skull uncracked, can itinerate Regent Street and perambulate the Parks, seeing the colours of thy heart waving on every fair head and fluttering round every cheek!

All the Year Round, September 1859

A particular fad for a colour began taking hold of Paris in the second half of 1857, and reached London the following year. The colour was mauve, the French name for the common mallow plant. As with most things fashionable, the Empress Eugénie appeared to lead the way. Emperor Napoleon III, easily intoxicated by pomp and display, married the 26-year-old Eugénie Montijo in 1853, and together they did their best to sweep away the dowdy parsimony of the Second Empire and restore the trappings of the grand court. The Emperor was a great leader by example, and he encouraged the Empress to promote trade by wearing heavy Lyons silks and all the *luxe* garments of Paris.

She needed little encouragement. Her own extravagant fashion sense, and her limitless budget, made her the target of every designer and court bulletin. This coincided with the publication of several new women's magazines devoted to cooking and clothes, and

Eugénie's every stitch was recorded with zeal. In Britain, the establishment of *The Englishwoman's Domestic Magazine* in 1852 by Samuel Beeton included hand-coloured fashion engravings, and ensured that trends from Europe would soon be knowledge in London, Norwich and Edinburgh (Beeton's wife Isabella published her book of *Household Management* in 1859 as an offshoot from the magazine).

Eugénie's particular fondness for mauve was first noted by the *Illustrated London News* towards the end of 1857, and it is possible that her preference influenced the choice of Queen Victoria's gowns for the marriage of her daughter the Princess Royal to Prince Frederick William in January 1858. Napoleon and Eugénie had visited the Queen and Prince Albert in London in 1854, during which the Queen had organised an appointment for Eugénie with her designer Charles Creed. But by the time Victoria and Albert paid a return visit to the Emperor and Empress on the occasion of the Paris Exhibition, Victoria's clothes were judged out of date, and it may have been that she now turned to Eugénie for some guidance.

For the royal marriage, the *Illustrated London News* reported that 'the train and body of her Majesty's dress was composed of rich mauve (lilac) velvet, trimmed with three rows of lace; the corsage ornamented with diamonds and the celebrated Koh-i-noor brooch; the petticoat, mauve and silver moiré antique, trimmed with a deep flounce of Honiton lace; the head-dress, a Royal diadem of diamonds and pearls.'

Three months later, the magazine noted that 'the mauve colour at present so highly fashionable is honoured by the especial favour of her Majesty . . . at the last levee her Majesty's train was of mauve velvet. The mauve is an exquisite shade of lilac. The mauve colour is also tastefully blended with black or grey.'

Inevitably, *Punch* did its best to deflate the new trend. The magazine had a particular distrust of most things fashionable and all things French, and managed to spear them both in a column entitled 'Imperatrice de la France et de la mode'. The world was indebted to the wife of Louis Napoleon for 'the endowment of that sumptuous and becoming colour which modistes and Mantallinis delight in calling mauve . . . We ask the ladies, the most impartial judges in the difficult art of personal adornment, if they can point their little finger to any other Empress whose edicts are more cheerfully followed by her millions upon millions of her admiring subjects.'

Initially, the mauve mania benefited William Perkin very little. The colour originated not from his aniline paste but from French supplies of murexide and purple dyes derived from various species of lichens. The lichen dye was of a brilliant colour, and like Perkin's purple, was capable of being produced in several strengths and shades. Writing in his *Manual of the Art of Dyeing*, published in Glasgow in 1853, James Napier noted that 'could this colour be obtained of a permanent character, and fixed upon cotton, its value would be inestimable'.

The lichen purple was produced primarily in Lyons by the firm of Guinon, Marnas et Bonnet, the company that had previously produced a strong yellow from picric acid. Initially it was applied only to silks and wool, but by 1857 a cotton mordant had been found, and the demand for the dye by far outweighed the firm's ability to supply it. This fact was not lost on Guinon's rivals, and in January 1858 representatives from the dye works Renard Frères and from several other competing firms travelled from Lyons to the London patent office in the hope of discovering the method of Perkin's aniline process. They also went to meet Perkin in Greenford Green, where they received a frosty reception.

In an attempt to maintain his monopoly, Perkin hurriedly prepared an application for the French patent office, and journeyed to Paris in April 1858 to finalise the process. It was here he was told that his application could not proceed, as he had waited more than 20 months since registering it in London, and the rules permitted a maximum of six. Unusually, this would soon work to Perkin's advantage.

Dr Crace-Calvert, the professor of chemistry at the Manchester Royal Institution, had already placed much of the process of making aniline coal-tar dyes in the public domain when he spoke at the Society of Arts in London in February 1858. When his speech was published, several French dye works seized on it and began aniline experiments of their own. While previously unable to produce purple by the same method as Guinon, Marnas et Bonnet (the firm which patented their process both in France and England), other companies turned by necessity to synthetic methods. And they found, after solving many of the problems that had hampered Perkin and Sons, that the colour they produced was brighter and more resistant to light and water than anything that had gone before. By the middle of 1859, Alexandre Franc et Cie and Monnet et Dury were both producing a magnificent and very popular purple. It is ironic that their colour was often called Perkin's purple or aniline purple or harmaline, and that Perkin and Sons would soon adopt the French name of mauve for their own shade. In Germany the colour was called aniline violet. In his contributions to the science journals, Perkin called his colour 'mauveine'.

Partly, William Perkin liked the name mauve because of its connotations with Parisian haute couture. As he wrote to his colleague Raphael Meldola, 'English and Scotch calico printers did not show any interest in it until it appeared in French patterns.' Thus the demand for Perkin's mauve was stimulated by his competitors abroad, but his British patent ensured that if a passion for it took hold in England his own rewards would be great.

Yet his success was not yet certain. Just as his colour was stirring hearts across the Channel, Perkin spent much of his time journeying to print works in Glasgow, Bradford, Leeds, Manchester and Huddersfield. Even those printers impressed with its fastness and intensity had to be instructed how best to use it. 'The dyers in those days had only been used to work with vegetable colouring matters and did not know how to apply basic colouring matters like the mauve . . . I had to become a dyer and calico printer to some extent,' Perkin explained. Thus at the age of twenty-one, Perkin left his factory and laboratory for weeks at a time to provide one of the earliest examples of 'technical service' – and he became an on-site expert and troubleshooter in a giant industry that had so recently been a mystery to him.

In this way he solved some of the last remaining dilemmas surrounding mauve – how best to apply it to calico and paper. He established new fixatives that would benefit the entire industry. When silk and cotton printers complained about the unevenness of his colour on their cloth, Perkin presented them with a new method of producing level dyeings by using dyebaths containing lead soap – and in this way soap baths gradually became another standard dyeing procedure throughout Europe. By the middle of 1859 his dye paste and concentrated mauve solution had been shipped in vast quantities not only to Thomas Keith in Bethnal Green, but also to the Scottish calico printer James Black and Co. in Dalmonach, Dumbartonshire; soon after, it would be in great demand in Leeds, Manchester and Bradford. In Perkin's phrase, 'They were clamorous for it.'

The traditional Glaswegian dyer who had doubted if anything would ever take the place of natural madder would then ask Robert Pullar, 'How did you come to think the thing was good?'

Pullar replied: 'I tested it and I felt convinced it was a good thing.'

'Eh,' the dyer said, 'ye must a' thought me a great fool that day.'

Soon after, the success of mauve was the subject of satire. At a Drury Lane pantomime a character remarked how now even the policemen told people to 'get a mauve on' (Victorians generally pronounced the colour 'morv'). Some amusement was had from the unhappy phrasing of a newspaper notice in July 1859: 'Found, on the 30th ult., a handsome Lady's Parasol, left there by two ladies, of mauve colour, lined inside with white, which may be had at Arthur's Stationery Warehouse . . .' It was the first time the word had appeared in *The Times*.

The following month, *Punch* wrote of a London in the grip of the Mauve Measles, an affliction 'spreading to so serious an extent that it is high time to consider by what means it may be checked'. The magazine suggested that doctors were arguing about symptoms and origin. 'There are many who regard it as purely English growth, and from the effect which it produces on the mind contend it must be treated as a form of mild insanity. Other learned men, however, including Dr Punch, are disposed rather to view it as a kind of epidemic, and to ascribe its origin entirely to the French. Although the mind is certainly affected by the malady, it is chiefly on the body that its effects are noticeable.'

Punch described the disease as infectious, beginning with the eruption of 'a measly rash of ribbons' and ending with the whole body covered in mauve. It detected that it was mostly women who were afflicted, for any symptoms in men could usually be treated 'with one good dose of ridicule'.

The most complete, and affectionate, account of what a colour could do appeared in September 1859 in *All the Year Round*, a new weekly journal. This publication was 'conducted' by Charles Dickens, and partially written by him; the journal, a successor to his weekly *Household Words*, contained the serialisation of *A Tale of Two Cities*. The author of an article called 'Perkins's Purple', in which the chemist gained an extra 's' throughout, is uncredited, but he certainly possessed a classical education.

'Let other men sing the praise of Hector and of Agamemnon, be it for me to sing the praise of Perkins, the inventor of the new purple.' The colour, 'which tradesmen foolishly call Mauve', made Tyrian purple look 'tame, dull and earthy indeed'. And Perkin's purple was so much better than the French variety which 'in waistcoats stained your shirts; in gloves, it gave you dyer's hands'.

The author believed that modern chemistry had several aims, including research for medical improvements, but by far the most significant was surely trade, 'and of the discoveries of our commercial chemistry Mr Perkins's discovery is one of the greatest and most brilliant . . . Alchemists of old spent their days and nights searching for gold, and never found the magic Proteus, though they chased him through all gases and all metals. If they had, indeed, we doubt much if the discovery had been as useful as this of Perkins's purple . . . A discovery that benefits trade is better for a man than finding a gold mine. It is, in fact, like this Perkins's purple, the key to other men's gold mines.'

Perkin, it was noted, was not like other inventors, in that he should soon became wealthy. With others, the author observed, 'fame comes, but when the money should flow in, there is a hitch, a frost, a blight.' But Perkin maintained the English patent, and thus the new colour had to be purchased from him directly. 'The Persian king, who offered a large reward to the discoverer of a new pleasure, by which he did not necessarily mean a new sin, would have buried Mr Perkins in a well full of diamonds. He would have pelted him to death with gold pieces, or have erected to his honour golden statues.'

The extent of mauve mania was documented in every detail; it was difficult to step into wealthy London without thinking there was something wrong with your eyes.

> One would think that London was suffering from an election, and that those purple ribbons were synonymous with 'Perkins for hever!' and 'Perkins and the English Constitootion!' The Oxford-street windows are tapestried with running rolls of that luminous extract from coal tar . . . O Mr Perkins, thanks to thee for fishing out of the coal-hole those precious veins and stripes and bands of purple on summer gowns that, wafting gales of Frangipanni, charm us in the West-end streets, luring on foolish bachelors to sudden proposals and dreams of love and a cottage loaf.

As I look out of my window, the apotheosis of Perkins's purple seems at hand – purple hands wave from open carriages – purple hands shake each other at street doors – purple hands threaten each other from opposite sides of the street; purple-striped gowns cram barouches, jam up cabs, throng steamers, fill railway stations: all flying countryward, like so many migrating birds of purple Paradise.

Mauve was the rage until 1861, and its prevalence was sustained by the other dominant craze of the day – the crinoline. The crinoline, the voluminous hooped iron cage that first swept down London and Parisian streets towards the end of 1856, represented the perfect advertisement for any new colour: you just couldn't miss it.

At least some of the credit for its success must go to the most prominent English couturier of the day, Charles Worth. Worth was born in Lincolnshire, worked for a while at Swan and Edgar, and then moved to Paris in the 1840s. His designs for the leading fashion house of Gagelin and Opigez, and later for his own firm of Worth and Bobergh, brought him into contact with several royal families, including Empress Eugénie. It was said that Eugénie had herself invented the crinoline to hide her pregnancy, but more likely it was Worth who developed the dress with the help of an English colleague. The crinoline has been described as the first application of the machine-age to women's dress; the massive steel structure reflected the glories of the Crystal Palace and monumental new bridges, and a great many were done up on newly imported American sewing machines. In 1859, it was said that Sheffield was producing steel wire for half a million crinolines each week.

Worth elevated dressmaking into dress designing, combining the new mathematical approach to tailoring with a flair for the ludicrous; he now wrapped women in the epitome of Victorian flash. The huge bird-cage structure of the crinoline eliminated the need for the many horsehair petticoats that had bulked out the ever-larger dresses from the start of the decade. Uniquely for a popular dress, the crinoline also caused many deaths. It was comfortable to wear – women liked to think of themselves as floating like a cloud – but it was dramatically cumbersome, not least in a period of increasing social mobility and popular rail travel. Madame Carette, a member of Empress Eugénie's court, wrote caustically of the dress that, when sitting, 'it was a pure matter of art to prevent the steel hoops getting out of place . . . To travel, to lie down, to play with the children, or indeed merely to shake hands or take a walk with them – these were problems which called for great fondness and much good will for their solution. It was about this time that it gradually went out of fashion for a man to offer his arm to a lady when he wished to accompany her.'

There were reports of women being engulfed by flames after catching their dress by an open hearth, and the Princess Royal burnt her arm after brushing past a candle. The worst incident occurred in a cathedral in Santiago, when up to two thousand women were burnt when a fire in the hanging drapes spread to their dresses. The dress inevitably invited other health scares: the journal *Once a Week* included an article called 'Dress and Its Victims', in which it warned that women would be at severe risk from the climate. 'Any medical man in good practice can tell of the spread of rheumatism since women ceased to wear their clothing about their limbs, and stuck it off with frames and hoops . . .'

In 1859, Empress Eugénie famously announced that she had given up the crinoline, but the news had little effect on its popularity, which declined only steadily over the next five years. Significantly for William Perkin and the textile trade, the dress appealed across all classes, and soon found itself to be not just a simple layer of silk or tulle, but elaborately sculpted from a great many tiered layers of additional flounces and ruches. In 1859, when the dress reached its largest circumference and consisted of perhaps four skirts and many trimmings, hundreds of yards of dyed material were needed for its construction. Dyemakers couldn't believe their luck: their order books swelled not just from the huge demand for dress material, but from the knock-on effects of newly-exposed ankles and the subsequent desire for coloured stockings and new petticoats.

By 1860, Perkin and Sons was meeting large export orders from Stuttgart, Amsterdam and Hong Kong. Initially the prices were high – Perkin obtained £6 for each litre of the mauve solution that was then diluted tenfold – and Perkin swiftly became rich. He began receiving medals from abroad in honour of his work. In 1859, he was sent a certificate from the Société Industrielle de Mulhouse, a region that once believed it alone led the world in new dyes. Accompanying the certificate was a silver medal and a letter from Dr Daniel Dollfus, the secretary of the society, praising the shade that had already given rise to so many applications 'and seems to promise even more'.

What it promised was a new way of looking at the world. A few months later, Perkin discovered that his method of making mauve was being copied throughout Europe, and was being used to make other colours to meet new demands for the latest fashion trends, particularly the new lines of walking dresses and women's sports costumes for croquet and tennis. Mauve, and the other aniline dyes it inspired, combined with new directions in fashion in ways that even the most ambitious dyemaker could never have imagined. No wonder *Punch* and traditionalists disapproved: for they were witnessing an early show of the female independent consumer.

Within two years of Perkin's invention it seemed that everyone was having a go at dyemaking. Industry had shown Victorian chemists what was possible, and now nothing seemed beyond achievement; an eighteen-year-old had created a new shade for a woman's shawl, and the full force of chemical ambition was unleashed. And of course there was much money to be made, and many fortunes to be lost, and a great amount of litigation.

Only two years after mauve had been the rage of Paris and London, its creator acknowledged that its best days were probably over. What people really wanted now was Verguin's fuchsine,

Manchester brown, Bismarck brown, Martius yellow, Magdala red, Nicholson's blue and Hofmann's violets.

There are great views from Don Vidler's midtown Manhattan office, but its occupant has grown familiar with them. In the middle of November 1999 he was more excited by the prospect of salesmanship. Vidler, a friendly man in his early forties, picked out a small pink T-shirt from a rack by his desk. 'This is from Banana Republic,' he said. 'Brand-new and just in the stores, a completely machine-wash/dry garment. Banana Republic has it, and Liz Claiborne, Next in the UK, Diesel, Dockers and Marks and Spencer takes a lot. If you go to the main store in Oxford Street they've got a ton of it in there and they usually promote the hell out of it.'

Vidler is talking about Tencel, the first new textile fibre developed for thirty years. He likes this product so much that he wears it himself – a grey Tencel polo shirt and brown Tencel cords. Vidler's clothes are really made from wood pulp.

Vidler works as the sales director for the fibre company Acordis, a company that spun off from Courtaulds, and is primarily responsible for the future of Tencel as a brand name. His job is to sell to the American mills, to persuade them to use the fibre in their woven and knitted fabrics.

Tencel has been in commercial production for seven years. 'I won't lie to you,' Vidler said. 'It's not yet on a level with DuPont's Lycra, but people are just starting to ask for Tencel by name.' The product is most successful with high-end women's sportswear, but it's big too in indigo denim. 'If you buy your pair of Levi jeans, you beat them up and wash them for a year and then they get soft the way you like them. Tencel is like that literally out of the box. What sells Tencel is the performance and a real soft hand [a peach-skin fuzz, soft to the touch].'

The story of the product begins in England, in Coventry, about twenty years ago. Courtaulds, the company that made a lot of its early money from a black dye (and sold it as being the blackest black you could buy), was making large quantities of rayon. But rayon production is a messy operation, with a lot of chemicals and much effluent. In an attempt to come up with a better rayon, the researchers at Courtaulds came up with something else.

Lyocell is produced from wood pulp through a solvent spinning process: the pulp is dissolved in an amine oxide and the result is forced into a water bath through fine jets. The solvent is washed out, and the fibre forms into fine filaments from which the staple Lyocell product is made. Rayon takes about 24–36 hours from wood pulp to fibre, but Tencel – a TENacity CELlulosic – takes only 2–3 hours, and 99 per cent of the solvent is recycled.

But Courtaulds discovered that there was a problem when it came to dyeing.

Tencel fibrillates like crazy – little pieces of the fibre stick out from the main body like hairs on an arm – and that is what gives its peach-fuzz appeal. If the fibrillation is not controlled very carefully as it goes around a dye jet in a metal drum it can get hairy and pilled and matted. Don Vidler says that it has taken a while for people to realise that if they want the true Tencel peach-touch it's a little bit more involved and costly than dyeing other fibres, but the results should be worth it. He takes out three indigo swatches, one Tencel, one cotton, one rayon, all placed in the same indigo dyebath, the first more lustrous than the others.

In November 1999, several leading fashion designers came in to Don Vidler's office to look at colour trends for Fall/Winter 2000 and Spring/Summer 2001. Tencel was nothing without the latest colours, but what would they be? And how would anyone know?

'We invited in all the stylists from Liz Claiborne, Donna Karan, really everyone in designer names,' Vidler explained. 'Also the next level down – the people that knock off the big names. Also the buyers for the big stores like Macy's and Bloomingdale's. They came in, and Sandy MacLennan flew in from his office in London and did his talk about colours and trends and what's going to happen.'

Sandy MacLennan brought with him some beautifully designed card brochures, each displaying a range of colours for the fashion seasons a year or eighteen months away. The colours did not actually have names, but they had moods. There was Trace, which ranged from beige to brown, and was described as 'warming . . . barely there . . . neutrals with pedigree . . . functional and exquisite'. Then there was Merge, which went from black to crimson, in which 'contrasts combine . . . dark and glimmering . . . deep, ravishing blends'. Next was Push, 'soft brights . . . solar glow . . . a surge of optimism . . . contemporary mix' (the colours were terracotta to grey). And finally Filter, from blue to grey again, 'casual and rational . . . shadowed mid-tones . . . modern and transitional'.

'I never know which comes first,' said Don Vidler, once MacLennan had gone. 'Does a colour become the hot thing because a colour expert says so, or is it because enough people told him about it first?

'There are clearly some key influences who go around the world, and I think Sandy is one of them. In my previous job I worked with a woman who was really considered the dean of the American colour experts. She would say that avocado will be the hot new colour next year, and because she said it, it was. She would do the colours for the automotive guys. They were looking five years ahead, and so she could say to the apparel people, "Ford are going to be making purple cars in two years," and so the apparel people played off that.'

Vidler paused for a minute, and conceded that some of this colour prediction thing was funny. 'I mean how many different names can you come up with for red? There's flame red, orange red – they'll play-off music – there's disco red, hip-hop red. And God forbid that Calvin's red looked like Donna's red or Ralph's red. Once a designer gave me the tip of a matchstick and said, "That's the red I want for this sweater."'

There is another thing that amuses Vidler about colour. He remembers that on every single day of his presentations, Sandy MacLennan wore a black T-shirt or turtleneck and black trousers. 'If you looked in his audience I guarantee you it would be the same. The New York uniform for women is black skirt or pants and a black Banana Republic spandex top and a black jacket. You go on the streets and you don't see a lot of flashes of colour. That kills me: you have these gurus saying, "These are going to be the hot new colours," but God forbid if they're caught in them. They're always going to let someone else wear their magenta and fuchsia.'

7
THE TERRIBLE GLARE

You want things just right when you're paying all that money, don't you? The Queen certainly does: as we discovered this week, she issues a list of royal demands before arriving at foreign hotels. She doesn't want the management to think she's fussy, you understand, but could they please make sure that any flower arrangements do not contain anything mauve (or carnations of any colour) . . .

A six-page Buckingham Palace memo reaches the press.
From the London *Evening Standard*, November 1999

In 1860 it was an exciting time to be a scientist. Many scientific societies had begun to publish journals, and their size increased each month. There appeared to be a theory for everything and a solution to as much. At every place where scientists gathered they liked nothing more than to suggest that science had got the better of nature, and nowhere did this belief carry more conviction than in the field of colour. In France, the leading dye firms were swiftly redefining the processes they had maintained for 300 years, believing that skilful atomic juggling would produce many more. In Germany, Liebig's students were ordering nitrobenzene and aniline by the kilo to make tests of their own.

But in London, not everyone had yet grasped coal-tar's full potential. In 1858, August Hofmann informed the Royal Society that he had produced 'a crimson colouring principle' as a by-product of an experiment combining aniline and carbon tetrachloride.

Yet he failed to exploit it, or even test whether it might be suitable as a dye. It appeared that the aniline he had requested from Perkin was not to make colour at all, rather for a purpose he still regarded as having a more justifiable claim on his time. But three years later his opinion had shifted so radically that by the time the British Association met in 1861, it was Hofmann, rather than Perkin, who was hailed as the hero of tinctorial science.

How did this come to be? Partly, it was down to public relations. Even after the discovery of mauve, Perkin remained a modest and diffident man. His speeches at scientific gatherings were formal and confined to precise technical matters. He did not seek publicity in the newspapers, and the pressures of his factory and the growing competition from France ensured that he had no time for self-exposure. Besides, he had recently become a husband and a father. In 1859 he married his first cousin Jemima Lissett, and they moved to rented accommodation in the Harrow Road in Sudbury. Their first son, also called William Henry Perkin, was born a year later, and their second son, Arthur George, one year after that. Then they moved to their own house, also in the Harrow Road, where Perkin again set up a small laboratory and restructured a large garden and play area for his children. This was not the time to consider claims upon posterity.

August Hofmann, however, had other motives. He had been responsible for inspiring Perkin's initial interest in aniline, and had taught most of the young English chemists who were just now entering dye works in London, Manchester and Leeds in an attempt to find new colours of their own. Though he never equated the pursuit of a colour with a scientific career, he would not have mistaken the growing prosperity that rewarded their inventors from 1860 onwards; even then, his main aim was to advance his academic reputation further than his bank account.

Hofmann was a more engaging orator than Perkin, and liked to talk on a grand and visionary scale. With no factory to run, he could attend all the scientific gatherings, and in the early 1860s he began to write passionate reports about the industry he had unwittingly set in motion. 'There are several [glass] cases which appear to excite in a more than ordinary degree the interest and admiration of the public,' he wrote of one chemical exhibition. 'In these cases is displayed a series of most attractive and beautiful objects, set in sharp contrast with a substance particularly ugly and offensive . . . gas tar.'

He wrote more like a travelling salesman than the country's leading professor. At the same exhibition he noted that the crystalline specimen of one dye reminded him of 'the sparkling wings of a rose-beetle'. A collection of silks, cashmeres, and ostrich plumes were 'the most superb and brilliant that ever delighted the human eye. Language, indeed, fails adequately to describe the beauty of these splendid tints. Conspicuous among them are crimsons of the most gorgeous intensity, purples of more than Tyrian magnificence, and blues ranging from light azure to the deepest cobalt. Contrasted with these are the most delicate roseate hues, shading by imperceptible gradations to the softest tints of violet and mauve.'

Hofmann attended the British Association meeting in Manchester in 1861, where its new chairman Professor Fairbairn presented the usual twelve-month catalogue of achievement. Chemistry was having 'a most direct bearing on the comforts and enjoyments of life'. The nutritional value of many foods could now be measured. Water was being analysed and purified as never before. The treatment of disease was improving rapidly through our comprehension of atomic theory. Fairbairn noted, however, that the greatest chemical developments of chemistry had been in connection with the 'useful arts'.

By this, the professor meant practical industry. 'What would now be the condition of calico-printing, bleaching, dyeing and even agriculture itself if they had been deprived of the aid of theoretic chemistry?' he asked. 'For example, aniline – first discovered in coal-tar by Dr Hofmann, who has so admirably developed its properties – is now most extensively used as the basis of red, blue, violet and green dyes. This important discovery will probably in a few years render this country independent of the world for dyestuffs. And it is more than probable that England, instead of drawing her dyestuffs from foreign countries, may herself become the centre from which all the world will be supplied.'

Such optimism would be undone within a decade. Indeed there were warning signs visible only a year later, when London hosted the International Exhibition of 1862. August Hofmann, writing his *Report to the Juries*, noted how far his science had progressed since the last Great Exhibition of 1851, and concluded with a further prophesy of continued English dominance. England would soon become the greatest colour-producing country in the world, he noted, and, 'by the strangest of revolutions, she may ere long send her coal-derived blues to indigo-growing India, her tar-distilled crimson to cochineal-producing Mexico, and her fossil substitutes for quercitron and safflower to China and Japan . . .'

Visitors to the exhibition were most excited by the first demonstrations of the safety match ('a match which could not be ignited by friction alone'). They were also impressed with the South East Gallery, in which several rows of cabinets held the latest samples of dyed cloth from Alfred Sidebottom of Crown Street, Camberwell, from Henry Monteith & Co. of Glasgow, and from John Botterill of Leeds. Robert Pullar of Perth displayed his latest umbrella cloths and dyed cotton. And in the middle of the hall was a large glass case labelled 'William Perkin and Sons', containing a pillar of solid mauve dye the size of a stove-pipe hat. The block was the product of about 2,000 tons of coal-tar, sufficient to print 100 miles of calico. One contemporary writer described it as 'being worth £1,000, the quantity of colour it contains being enough to dye the heavens with purple'.

Not all visitors were so impressed. The French historian Hippolyte Taine visited London for the exhibition and found both the exhibits and the visitors who gazed upon them to be gaudy and unrefined. 'The exaggeration of the dresses of the ladies or young girls belonging to the wealthy middle class is offensive . . . gowns of violet silk with dazzling reflections, or of starched tulle upon an expanse of petticoats stiff with embroidery . . .' Walking on a Sunday in Hyde Park, he saw more bright colours than he had ever seen gathered in one place. 'The glare,' he observed, 'is terrible.'

Perkin's success with mauve had brought twenty-eight other dye-making firms to the International Exhibition. There were eight other companies from the United Kingdom, twelve from France, seven from Germany and Austria and one from Switzerland. Several of them showed hues that appeared to be exactly the same as those of their competitors; but the names were novel, and their brilliance and claims on fastness would have been unimaginable even five years before.

The chronology of the new colours began in 1859 with a former French schoolteacher. Emmanuel Verguin had worked for a while as the director of a factory which made yellow from picric acid, and he knew that a new colour was a valuable commodity. In January 1858 he joined another firm, Renard Frères & Franc, one of whose workers had recently visited Perkin and Sons in the hope of obtaining trade secrets about aniline. Despite his previous experience, Verguin appears to have signed a highly restrictive contract with his new employers, guaranteeing them the rights to any new colour he might discover in return for one-fifth of any profits. Within weeks of joining, his researches into aniline had produced a rich crimson red, probably a very similar shade to that discovered by Hofmann. He called the colour fuchsine, after the flowering shrub fuchsia, and by the end of the year the colour was in demand from Cherbourg to Marseilles.

Fuchsine was produced in far greater quantities than mauve, being used first for military uniforms and then widely as the latest crush in fashion. In Britain it became known as solferino and then magenta, taking the names from the Franco-Piedmontese war against Austria and Garibaldi's victory in North Italy, where the dye matched the colour of the soldiers' tunics. The demand for the colour was such that it was produced not only in Lyons, but soon, with the slightest manipulations of molecular formulae sufficient to bewilder any patent office, in Mulhouse, Basle, London, Coventry and Glasgow. In Britain it was first used on a large scale in Bradford by Ripley and Son, who recorded with pride how they were first offered the dye by Renard Frères in February 1860 at £5 per gallon, but just a month later had struck a deal for £3 per gallon. In time, Renard Frères set up its own manufacturing base in Brentford, Middlesex.

In east London, magenta made fortunes for the firm Simpson, Maule and Nicholson, a company that had only recently transformed itself from producing chemical supplies such as aniline and materials for photography into a fully scaled-up dye works. Edward Nicholson, yet another of Hofmann's former pupils, formed an informal partnership with his old tutor, providing him with some magenta crystals. Hofmann then set about analysing the new dye's molecular composition. He changed its name once more – this time to rosaniline – and his work unlocked the hidden constitution of almost all the new aniline dyes.

From this applied approach any number of new colours could be constructed. Within months, industrial chemists were able to produce quite a pattern book: tiny structural tweaks turned the magenta to aniline yellow, then to bleu de Lyon, bleu de Paris, and Nicholson's blue. A little later there would be aldehyde green. Hofmann himself produced two new shades of violet.

Such developments had not gone unnoticed at Greenford Green. Perkin observed the gradual decline of his original mauve with some disdain, and set about making improvements. Thus he made dahlia, an intermediate between mauve and magenta, and the first industrial supply of aminoazonaphthalene, the colour he had first produced with his friend Arthur Church while still at the Royal

College and which he could now dilute from scarlet to orange; for reasons of nomenclature alone, this last was not an easily marketable item. Perkin discussed these new colours before the Royal Society, and the contrast with Hofmann's excited prose could not have been more stark. Perkin, still only twenty-three, remarked later that perhaps his greatest success in these talks was the fact that they had been attended by Michael Faraday.

Perkin was aware that he was losing his early lead in aniline dyes, and appeared to care little. He contributed a steady stream of articles to the journal of the Chemical Society, many on subjects other than dyestuffs. In September 1860 he obtained great satisfaction from a letter sent by the director of scientific studies at the École Normale Supérieure in Paris. In this he was informed that the contents of a small glass phial he had supplied had been analysed as pure paratartaric acid, then regarded as a novel and important synthesis. The letter continued: 'I should be very grateful if you could send me a portion of the succinic acid which you used to prepare the paratartaric acid. I am very keen to investigate whether, by any chance, it might be endowed with the action for polarised light.' Perkin's correspondent was Louis Pasteur. 'Please forgive my indiscretion,' Pasteur wrote. 'All this is so important from the point of view of our ideas on molecular mechanics that I cannot contain my impatience to know. I am also very happy that this occasion has provided me with the pleasure of entering into relations with one of our best chemists.'

As the decade wore on, Perkin's newer colours enjoyed more success. Britannia violets, made by heating a solution of magenta with turpentine, produced deep bluish shades. There was also Perkin's green, popular for a while in calico printing, and aniline pink, and Perkin's own form of magenta, involving the use of mercuric nitrate, which soon had to be curtailed due to the harmful effect of the mercury on his workmen. Perkin also made the salts and copper compounds necessary for aniline black, and greatly improved the methods of dyeing wallpaper. Regulars at the Black Horse pub near the Perkin factory would observe how the Grand Junction Canal turned a different colour every week.

The new colours had predictable effects – an increase in the British, German and French balances of trade, and many tingles of excitement in the fashion world. Some effects were clearly detrimental to the traditional trade of natural dyestuffs. In 1862 August Hofmann observed a stark difference in the price and demand for cochineal, safflower, indigo and madder from that just three years earlier. Imports of scarlet cochineal, for example, had increased by more than 50 per cent in weight between 1847 and 1850, but all had changed with the advent of aniline dyes. From a peak of 15 francs per kilo in 1858, the price had fallen to 8 francs. Within two years the price of safflower had fallen from 45 francs to 25, and picric acid had all but curtailed the supply of natural yellow dyewoods. Even indigo, which had not yet been synthesised, was no longer used in silk dyeing, replaced by the artificial blues and violets. 'Thus,' one contemporary French writer regretted, 'of three dyeing materials hitherto considered indestructible elements of the commercial prosperity of tropical countries, we find indigo diminished in its applications, and cochineal and safflower very notably depreciated, solely and exclusively by the work of the chemist.'

And then there were less expected results, including a show of begrudging respect towards the industry from those who had previously only ridiculed the discovery of mauve. *Punch* appeared to do a complete about-turn, although there was hyperbole in its comic verse:

> There's hardly a thing that a man can name
> Of use or beauty in life's small game
> But you can extract in alembic or jar
> From the 'physical basis' of black coal-tar –
> Oil and ointment, and wax and wine,
> And the lovely colours called aniline;

You can make anything from a salve to a star,
If you only know how, from black coal-tar.

More important, perhaps, was the impact that the success of the dye trades had on the recruitment of young chemists. Ten years after Perkin's discovery of mauve, organic chemistry was perceived as being exciting, profitable and of great practical use. Perkin and Sons employed the talented chemist Charles Greville Williams (later to make his name in the field of hydrocarbons and secure four new colour patents of his own), and many others attached themselves to the huge dye works springing up throughout Europe. In Britain, it appeared that about two-thirds of former pupils of the Royal College of Chemistry later worked for dye firms such as Simpson, Maule and Nicholson, or Brooke, Simpson and Spiller, or Read, Holliday and Sons, and most of these were later honoured with fellowships of the Royal Society. There had been no finer example of the mutually beneficial relationship between science and industry, and these new minds would soon unlock the structures of a great many other carbon compounds with direct benefits for medicine, perfumery and photography.

What the young chemists had not expected from the dye trade was an education in litigation, but many soon found themselves embroiled in bitter disputes over patents. This was bound to happen. All aniline dye companies soon found that they would make most money by imitating the most popular new tints.

The problem for the courts was that the colours often looked exactly the same; the more skilled the dyers in copying a patented recipe, the harder to distinguish between, say, six shades of blue. There was no colour chart to which experts could refer and make distinctions, and molecular analysis was still in its most primitive form. In addition, there were interminable disputes over process – the extent to which the same shade, derived from a slightly different method of manufacture, contravened an existing patent.

The hottest arguments inevitably concerned the colours in greatest demand. In the early 1860s this meant magenta. The most prominent lawsuits came from Lyons, where Renard Frères & Franc sought to protect their fuchsine by successfully suing companies in Mulhouse and Paris. These cases outraged Renard's rivals, and in 1861 over 100 Lyons firms petitioned the French Minister of Agriculture and Commerce to establish an independent commission to resolve the claims surrounding fuchsine and other aniline dyes.

They obtained little satisfaction, for at the end of 1863 Renard Frères & Franc entered into an alliance with Crédit Lyonnais to set up La Fuchsine, a huge monopolistic alliance formed out of multiple mergers with several leading dye companies. This alliance forged close links with Simpson, Maule and Nicholson in London, which, in return for a licence to manufacture magenta, granted the French company licences for Hofmann's violets. La Fuchsine was also grateful to August Hofmann for another reason: it was his expert testimony which influenced the French courts to uphold its claims of patent infringement.

La Fuchsine ended in failure a decade later, not least because of poor management and several claims that its arsenic acid process was poisoning the locals. In one case, the wife of a signalman died very close to one of its factories, and a post-mortem established the presence of arsenic in her organs. The same type of arsenic was detected in the well from which she drew her drinking water, and in all the wells and subsoil within 200 yards of the fuchsine factory. The company paid out to her family, and, following public protests, production of the red dye ceased.

In England, many of the earliest disputes concerned attempts by Henry Medlock and Simpson, Maule and Nicholson to protect the very same arsenic acid oxidation process for magenta that had caused such concern in Lyons. Some £30,000 had been spent defending the process, as the London company actively sought out offenders. Two big cases occupied the London High Court for several months, and both centred on the precise interpretation of the term 'dry' arsenic acid. At one, the judge told the jury that he was relieved that they, and not he, had to decide on the case, such had been the bewildering and conflicting array of expert evidence. In both cases the patent was upheld. One losing party was obliged not only to pay Simpson, Maule and Nicholson damages and costs, but also to take out a series of public apologies in *The Times*.

William Perkin's patent wars were recorded in a trade journal, *Chemical News*. This reported the successful battles against British and French firms who were marketing imported aniline purple or making their own Britannia violets or Perkin's green by a similar process. These companies were fined several hundred pounds, and also forced to make humiliating apologies. But Perkin observed that he could have spent his entire career in the courts, such was the money to be made from synthetic dyes. In 1865 the gross profit made by Perkin and Sons stood at about £15,000, even though the price of dyestuffs was by then falling sharply.

In his last years, Perkin would speak out against the inadequate system of legal protection of his work. At the time, he noted the increased number of overseas visitors to his factory, particularly Frenchmen, and the growing amount of German and Polish-born chemists employed at various British dye firms. In Manchester, Roberts, Dale and Co. employed the calico printer Heinrich Caro, who made important technical advances, and Carl Alexander Martius, who patented Manchester Brown and his own brand of aniline yellow. A man called Otto Witt made new dyes at a firm in Brentford, north of London. Peter Griess, employed as a brewer in Burton-on-Trent, developed an important new family of colours, the azo dyes. And Ivan Levinstein set up his own dyemaking company in Salford, and made blues and oranges.*

* Griess was particularly useful to a Burton firm of brewers because he had helped them discover an important ingredient that imparted a desirably bitter taste to beer and for years had given an advantage to London brewers. The ingredient was picric acid, the yellow wool dye.

With the exception of Levinstein, all the chemists would return home with their knowledge, and they would make their fortunes (Martius was co-founder of the predecessor of the dye company AGFA). But none would be as missed as August Hofmann. His departure in 1865 was believed to have been inevitable ever since his main sponsor Prince Albert had died four years before; from Germany, Hofmann would speak of the loss of financial support and lack of encouragement for his teaching. He criticised the British government for failing to grasp the importance of chemistry both as a pure science and as a means to further industrial advance. His forecasts from the International Exhibition of 1862 he now saw as being rather optimistic. He was lured back to Berlin by the promise of a lucrative research post and large laboratories that he could design himself. It was with much genuine disappointment that he noted that England had no such ambitions. Hofmann was given three years of absence from the Royal College to pursue his career in Berlin, but he never returned.*

For William Perkin, too, it was a time of farewells and sad depar-

* Before he left for Berlin, Hofmann received a cheering letter from Jaz Clark, a secretary at Windsor Castle. It was dated 27 March 1865: by this time it was clear that his presentations had advanced from broad theoretical discussions of the great potential of inorganic chemistry to practical demonstrations of the miracle of synthetic dyes.

> Dear Dr Hofmann,
>
> I have received the Queen's commands to express to you the great pleasure which Herself and the Royal Family derived from the very interesting and clear lectures on chemistry, and the beautiful experiments by which they were illustrated, delivered in Windsor Castle last week.
>
> Her Majesty also admired the numerous beautiful specimens of richly coloured silks and wools, the results of the recently discovered aniline dyes, and perceives clearly the great advantage to the material interests of this country which must result from the discovery of these beautiful colours, and it gave her great pleasure to learn that they originated in researches conducted in the Royal College of Chemistry, in which his late Royal Highness the Prince Consort took so much interest . . .

Within a few years, the advantage of the 'material interests' mentioned in the letter would have fled with its recipient to Germany.

tures. His father died in 1864, and the factory he had financed closed for several days as a mark of respect. Perkin grieved alone, for his wife had also died of tuberculosis a few years before.

He may have taken some comfort from the fact that mauve had transformed itself from the colour of frivolity and display to the colour of mourning. Before her marriage to the Prince of Wales in 1863, Princess Alexandra made a grand entrance into London in a half-mourning dress of pale mauve poplin, and Queen Victoria would graduate from black to mauve within four years of losing her darling Albert.

By 1869, however, mauve was all but forgotten. Its replacement – another great new colour, another rage throughout Europe – rejuvenated the fortunes of the British dye trade. William Perkin was again responsible for its manufacture, but its success would lead to his acrimonious departure from the industry and a promise never to return.

'You can tell when a colour is taking off. You can spot it all over the place. You watch what people do when they shop, and what catches their eye first is usually the colour. Then they feel it, and if it still works for them they might try it on. There has to be a little bit of theatre, and some pleasure.'

Sandy MacLennan was back from his trip to Manhattan. In his East Central Studios in Shoreditch, east London, he was already thinking about Fall/Winter 2001, about the way he may edge us towards burnished neutrals or colours at the edge of sleep.

MacLennan, forty-seven, from the west coast of Scotland, did not always think of colours in this manner; once he thought such talk was absurd.

'There is only a small community that regards colour in this way,' he concedes. 'The design community don't need a guide book to get through this . . . we see colour as a catalyst for new ideas. It's the beginning point in any cycle.'

He says that each season's new colour forecasts begin with looking at the last – he's not simply dropping in on a season and inventing it. 'Then you look at the things that need to be there – the things that people will wear. You watch everything, talk to people, you see what people are wearing and what's selling. Then you build up this balanced offer.'

He knows that people wonder about his unique vocabulary. For Spring/Summer 2001 we will be Dreaming, Idling, Exploring and Flirting, and more specifically touching new surfaces and being softly greened with bone-dry basics. Blues, greens, organic ochres and crimsons.

This palette is again for Tencel. 'You want to try and sex it up in a way that is completely relevant for that product,' MacLennan says. 'You wouldn't want to use words that were all to do with transparency or glass or plastic, because that's another world, and Tencel is a very natural world.'

When he presents his new colours to a company (MacLennan also forecasts and designs for DuPont, British Home Stores, Nordstrom and Liberty), he never suggests that they have to do it all. 'We simply try to talk about colour in an emotional way. Like the way people describe wines – in terms of what it evokes – what emotions, what memories, what aspirations.'

MacLennan selects one of his colour brochures. 'It's not easy to say which colours are going to be the ones – but it won't be that [pointing to yellow] because yellow is a terribly difficult colour to sell. But that lime-green and this orange thing is still going strong – that's three years it's been going. Over winter, consumers forget about these colours, forget that they enjoyed this colour on the beach, and so when spring comes round they find it charming again.'

In the winter of 1999, MacLennan detected a liking for a family of colours including purple and grape and blood-reds. The other trend over the last few years has been the growth of green. 'For the longest time people thought of green as German – oh, we don't wear green, green is difficult and always a problem. But six years ago there was a change in people's perceptions. Blue took a bit of a back seat, coinciding with people deserting denim. The tones tend to go clearer and more acidic in the summer, and richer and heavier in the winter.

'Mauve will be forever associated with the seventies – Biba and all that. It enters the psyche in a sensual way, and is often perceived as decadent or forbidden. Doesn't suit everyone. People often buy it in error and then are alienated from it for good. In the family of purples, I always think of mauve as something lighter, greyer, softer than all its friends. Culturally it can be quite awkward globally. People in Japan won't wear it because they associate it exclusively with royalty. You can't use it as a fashion statement very easily over there. And there's that whole cardinal thing as well. When it does arrive, it comes and goes quite quickly. Mauve tends to swamp the things it comes into – if you add it to other colours it just tends to go mauve, rather than a gentle mix. I think it has a mythical feel. We're moving on to describe colours by smell, and I think mauve would be incense, probably patchouli oil or a smoky, holy, churchy smell.'

Sandy MacLennan is aware that 70 to 90 per cent of all sales are always black, all the year round. Followed by grey and navy. But he doesn't find this soul-destroying. For him, the key is to pick upon the things that are going to be the other 10 per cent.

'I think we're used by companies to give them an edge. There's too much fibre in the world, and therefore potentially too much fabric. There's too much product being made, and the choice for the consumer is so big. Even the slightest edge can make a big difference. The luxury garments – luxury wool or silk – will always sell itself, but how do you sell the crap? There is always going to be too much crap around, so the companies that sell it have got to be quite busy with their marketing and understanding of what people want. Years ago it was very different. In the nineteenth century, people made something and you sold it.'

8

MADDER

The local spy – and there was one – might thus have deduced that these two were strangers, people of some taste, and not to be denied their enjoyment of the Cobb by a mere harsh wind . . . he would most certainly have remarked that they were people of a very superior taste as regards their outward appearance.

The young lady was dressed in the height of fashion, for another wind was blowing in 1867: the beginning of a revolt against the crinoline and the large bonnet. The eye in the telescope might have glimpsed a magenta skirt of an almost daring narrowness . . .

The colour of the young lady's clothes would strike us today as distinctly strident; but the world was then in the first fine throes of the discovery of aniline dyes. And what the feminine, by way of compensation for so much else in her expected behaviour, demanded of a colour was brilliance, not discretion.

John Fowles, *The French Lieutenant's Woman*, 1969

Ordinary women always console themselves. Some of them do it by going in for sentimental colours. Never trust a woman who wears mauve, whatever her age may be, or a woman over thirty-five who is fond of pink ribbons. It always means they have a history.

Oscar Wilde, *The Picture of Dorian Gray*, 1891

Regarded from any angle, the madder plant has never been an attractive specimen. William Perkin examined it in 1868, and found leaves rough with prickles. *Rubia tinctorum L* is an herbaceous perennial that grows up to five foot high. The flower, such as it is, is small and greenish yellow, its stem square and jointed, its root cylindrical and fleshy. It is propagated from suckers, and the roots give off a strong odour.

Only a dyer could love such a plant. Up until 1868, dyers loved it very much, for madder provided half the world with red. Even one decade after the discovery of the coal-tar dyes, annual imports of madder to the United Kingdom (principally from France, Holland, Turkey and India) were valued at £1 million. Alongside indigo, madder had provided the textile dyer with a staple ingredient of their trade for centuries. Between 1859 and 1868, wool and calico printers imported an average 17,500 tons of madder and its derivatives each year, much of it used in Scotland by the likes of Reid and Whiteman of Maryhill, and James Hendrie of Arthurlie, and J. & W. Crum and Co. of Thornliebank. Among many things, madder dyed soldiers' trousers.

In 1876, William Perkin planted some madder on a piece of land opposite his home in Harrow. It was a symbolic act from a man not much given to such displays: he was growing it, he said, 'lest the breed should become extinct'.

By 1876, the import of madder into Britain had fallen to 4,400 tons, a quarter of the intake a decade before. Its price had fallen by almost a third in the same period, to £1 per hundredweight. The decline showed no sign of abating; madder growers, once secure in a lifelong industry, were witnessing their world collapse around them. They could blame the advances of chemistry, and, if they had known his name, they could have specifically blamed William Perkin.

Madder is indigenous to western Asia, and was introduced to Spain by the Moors. It was widely used in Holland in the sixteenth century, and in Avignon a century later. Its use as a dye, on textiles and occasionally ceramics, has been described by Pliny the Elder and in the Talmud. In 1809 the French chemist J. Chaptal detected a pink pigment he had found in a shop in excavations at Pompeii that may have come from a madder lake. In 1815 Humphrey Davy found a similar colour, probably from a similar origin, on a broken ceramic vase at the Baths of Titus. In the seventh century, textiles dyed with madder were sold at St Denis near Paris, and Charlemagne is known to have encouraged its growth. Vegetative remains of madder were found in excavations from the tenth century in York. As recently as 1993, the conservationist J. H. Townsend identified madder lakes in palettes used by J. M. W. Turner.

But chemistry believed that the Turners of the future would paint just as well with acrylics. On 8 May 1879, Perkin addressed the Society of Arts in London on his efforts to eliminate the need for ancient madder and synthesise its tinctorial properties in a factory. There was little remorse for the erosion of an age-old natural trade, but he did exhibit a great understanding of the traditional methods of cultivation and dyeing.

'The time from planting until the roots are drawn is from eighteen to thirty months,' he told his audience. 'When dried, the roots lose their reddish yellow colour and become of a pale red shade. The process of drying is conducted in the air or in kilns. When dry, the roots are beaten to remove sand, clay and loose skin . . .'

Depending on the mordant and strength of dye, the textile shades ranged from pink, red, purple and black. Another tint, a brilliant and strong red known as Turkey red or Adrianople red, required a cotton mordant of olive oil, and the addition of a little animal blood.

But what of the colouring matter within madder which rendered it so desirable? Little was known of it until 1827, when two Frenchmen, Colin and Robiquet, introduced a method of producing garancine, a concentrated form of natural madder, and at the same time detected precisely what it was that gave the root of the plant its value. By heating ground madder in a test tube with various acids and potash they obtained a yellowish vapour which crystallised into bright red needles. They called this substance alizarin, from the Levantine term for madder, alizari.

Alizarin made up only about one per cent of the madder root, and colour chemists reasoned that – as with magenta and cochineal – finding the correct molecular constitution for alizarin would yield a far greater purity and efficiency, and some riches. A Manchester-born calico printer named Henry Edward Schunck spent much of the 1840s and 1850s investigating the madder root, believing that any artificial formulation would require a similar naphthalene base as previous synthetic colours. Like Perkin in his early days, his chemistry was largely empirical, and still involved an often fruitless trial-and-error process of adding or subtracting a variety of carbon and hydrogen elements and hoping for success.

But the nature of molecular understanding was changing rapidly. In 1858, Friedrich August Kekulé had explained the concept of isomerism (the presence of compounds with the same molecular formula but different arrangements of the atoms within the molecule). Yet Kekulé's groundbreaking *Theory of Molecular Structure* failed to explain the behaviour of benzene, the hydrocarbon present in coal-tar and known as an 'aromatic' compound because of its presence in scented oils. The answer appeared to him one evening while he was dozing in front of a fire, a vision of gambolling atoms in snake-like motion. 'One of the snakes had seized hold of its own tail, and the form whirled mockingly before my eyes.' In this way he reasoned that carbon atoms formed rings: his benzene structure consisted of six carbon atoms, each attached to a hydrogen atom (C_6H_6)

The analysis of chemical structures would now proceed on a more rational basis. Hofmann had already been able to define the chemical formula of fuchsine, and now was able to resolve its precise elemental formula. Two other chemists who made full use of this new knowledge also worked in Berlin – Carl Graebe and Carl Liebermann. In 1868, they demonstrated that alizarin did not have a naphthalene base but one of anthracene, another aromatic compound present in coal-tar. Graebe and Liebermann showed that anthracene consisted of three fused benzene rings, and they were thus able to synthesise alizarin in the laboratory – the first time a vegetable colouring matter had been made in this way – and they patented their discovery in December 1868. Unfortunately, their vastly complex method involved the use of prohibitively expensive and volatile liquid bromine in small sealed tubes, and they were unable to produce alizarin on any scale. To this end they enlisted the services of Heinrich Caro, the skilled industrial chemist who had returned to his homeland in 1866 after obtaining great practical experience in Manchester. Caro had returned to study and teach at the University of Heidelberg, but soon landed the job of director of research at BASF, Germany's largest dye factory. No one employed at these works underestimated the value of being able to produce alizarin in giant vats; demand appeared to be insatiable. But they soon discovered a rival.

William Perkin was ideally placed to exploit the situation. Not long after he had joined the Royal College of Chemistry in his teens, August Hofmann had instructed Perkin to prepare and examine anthracene from coal-tar pitch. As Perkin remembered at a memorial after Hofmann's death, he was perhaps 'more fully prepared than any other chemist of the day to appreciate the discovery of the relationship of alizarin to, and was naturally impelled at once to adapt it to practical requirements'. With mauve no longer in vogue and the price of magenta tumbling due to widespread competition, the possibility of adapting his Greenford Green works for the large-scale manufacture of alizarin presented a breathtaking challenge. Alizarin was not prone to the vagaries of fashion; unlike mauve or some of the aniline dyes that followed it, dyers did not need persuading of its usefulness or instruction as to its application. In Scotland alone there were more than seventy dye works keen to take out orders.

Within a year, William Perkin and his brother had devised two processes by which alizarin might be manufactured, and neither required the use of bromine. Initially, Perkin & Sons (the plant retained the name after their father's death) experimented with the original anthracene left over from Perkin's student days, and began a series of combustions. The most successful entailed heating anthraquinone (another aromatic benzene-like compound present in coal-tar) with sulphuric acid, the product then fused with caustic alkali. 'To my delight [the result] changed first to violet, and then became black from the intensity of its colour. On dissolving the melt, a beautiful purple solution was obtained, which gave a yellow precipitate when acidified, and on examination was found to dye mordanted cloth like garancine.'

As he did with mauve, Perkin sought the opinion of an experienced dyer, again in Scotland. On 20 May he sent samples to Robert Hogg in Glasgow, an expert in madder, who told him that the quality was superb. Perkin then sought a patent, noting later how 'this process has proved the most permanently important one yet discovered.' He began to gear up for mass production.

With their earlier colours no longer in great demand, the Perkins found they could adapt some of their existing apparatus for alizarin, but they also needed to expand. A new plant was planned, and new staff hired. Again, they faced many supply problems of key ingredients, not least with anthracene; tar distillers had previously found no use for it and failed to produce it. The first quantities were made on site by distilling tar pitch, but, as Perkin later recalled, it was necessary for his brother to visit 'nearly all the tar works in the kingdom' to show distillers how to separate anthracene. Other problems – the purification of the anthraquinone, the transport from Germany of fuming sulphuric acid – required the Perkins to alternate between two different production processes, and revert to some unorthodox methods. The compounds were combined in a large bath, before being placed in a revolving cylinder into which cannon balls were added to improve mixing. Another difficulty concerned pricing. Unlike many of the previous dyes, for which there was no direct natural competitor, madder had a fixed and competitive cost structure, and Perkin & Sons had to ensure that its production was of such a scale and efficiency that they could undercut the present price levels of 50 shillings per hundredweight for madder and 150 shillings per hundredweight for garancine.

But then they were hit by another problem, something that threatened the entire business. Perkin's patent for his first alizarin production process was granted on 26 June 1869, but the patent application for a very similar method had been filed just the day before by Graebe, Liebermann and Heinrich Caro of Berlin. Initially, Perkin was devastated, not least because Caro was a friend of his.

Some years later, Perkin would cite the inexact and lumbering workings of the British patent office as a great hindrance to industrial progress, but at the time of the alizarin dispute he limited his public frustration to a footnote on a paper he delivered to the Society of Arts. He noted that Graebe and Liebermann claimed that Heinrich Caro had formulated the process before Perkin, and that 'if any particular importance is attached to dates, the advantage rests without dispute with Caro, Graebe and Liebermann, for the filing of the patent . . . was delayed through irregularity. The signatures had already been given in to the Patent Office, Berlin, on the 15th June.' Perkin fumed: 'I may remark, in reference to the first statement, that Graebe and Liebermann neither give or adduce any evidence to substantiate their claim to priority.' He confirmed that his dyed pattern samples had been sent to Robert Hogg several weeks before the dates in question, and that his patent had also been delayed. 'Therefore, their conclusions, from the argument as to dates, should be reversed . . . Without wishing to detract from Graebe and Liebermann's original discovery, we may say that the birthplace of the manufacture of artificial alizarin was in England.'

In the end, both patents were granted. The BASF patent was registered one day earlier, but Perkin's was sealed first. Perkin invited Heinrich Caro to Greenford Green to carve up their markets, and it was agreed that Perkin & Sons should hold the British monopoly for several years while BASF controlled the market in mainland Europe and the United States. The first shipments of artificial alizarin left Harrow on 4 October 1869, and by the end of the year one ton of paste had been made. Production at BASF had been held up due to the Franco-Prussian war, and Perkin found that he had the world market to himself for almost a year. 'In 1870 we produced 40 tons,' Perkin noted. 'In 1871, 220 tons; in 1872, 300 tons; and in 1873, 435 tons.'

In the first year of production, William Perkin received some correspondence which suggested that he had not made alizarin at all; it was claimed that the fiery qualities of his colours had never been produced from natural madder: 'The red shades were more brilliant and more scarlet, and the purples bluer; the blacks were also more intense.' In May 1870 Perkin felt impelled to demonstrate to the Chemical Society that alizarin could indeed be separated from his commercial product, and that it possessed the same dyeing properties as the natural root.

The financial rewards were great, despite tumbling prices. Perkin & Sons received approximately £200 per ton. By the end of the 1870s this had fallen to £150 per ton, or about one-third of the price of natural madder in 1868. In the two years before June 1873, Perkin & Sons made an annual profit of about £60,000. William Perkin's personal fortune was approximately £100,000, precisely the figure Victorian businessmen used to denote a man of substance.

At the height of his achievement, Perkin addressed the Society of Arts about the 'wonderful' and staggering growth of the coal-tar colour industry. Omitting his personal role in the enterprise, he spoke of how the industry had 'acted as a handmaid to chemical science, by placing at the disposal of chemists products which otherwise could not have been obtained, and thus an amount of research has been conducted through it so extensive that it is difficult to realise, and this may, before long, produce practical fruit to an extent we have no conception of.'

But the fruits would not reside predominantly in Britain. In 1878 the estimated value of coal-tar production in England stood at £450,000, compared to £350,000 in France and Switzerland, and £2 million in Germany. This was only the beginning of Germany's dominance. A year later, there were seventeen coal-tar colour works in Germany, compared with five in France, four in Switzerland and six in England. Perkin & Sons was no longer among them.

In October 1999, a 47-year-old man called Robert Bud sat in his office at the Science Museum in London, and explained how he had once designed the museum's modest display on artificial dyes. A large, roundish man with short brown curly hair, Bud had been at the museum for twenty-one years and was now Head of Research (Collections). He also had a responsibility for organising conferences: his computer hummed with e-mails about a conference he was planning for July 2001 on Victorian culture. By July 2001 it would be 150 years since the Great Exhibition, 100 since the death of Victoria herself. He produced an early leaflet about the conference and said of its colour: 'It was the closest we could get to the original mauve.'

His office opened onto a stairwell, and from this another door led directly to the chemical industry section of the museum. It was shortly after 9 a.m.; the place was not yet open to the public, but the working displays were already grinding away. Dr Bud walked a few yards to the exhibits about dyes, a series of models and cases and backboards he helped construct thirteen years ago.

He touched a huge wooden dye vat from an ICI plant in Blackley, near Manchester, which until 1980 was used for making detergent. It still had that smell. 'Sulphuric acid was added to fish oil and salt water,' he said. 'Very crude, very simple, very typical. The chemistry was very sophisticated, but the technology was often not.'

There are four elements of Bud's dyeing display: the vat, three mannequins wearing clothes from the 1880s, 1930s and 1960s, a portrait of Perkin with a model of a man called Rudolph Knietsch (one of the developers of synthetic indigo) in front of him, and a glass case opposite containing bottles of powdered dye and a salesman's wool sample books. There ought to be something else, a small corked glass phial with the inscription: 'Original Mauveine prepared by Sir William Perkin in 1856'. But a sign says it is in Osaka, and should be back by March. When it returns it will probably go into the big new gallery on the history of technology on the ground floor.

'Dyes drove so many industries that it's easy to forget that they totally changed the way the world looked,' Dr Bud said. 'Before the synthetic dyes, you could argue that rich patterns of exotic colours were an elite thing. After them, everyone lived with such colours as mauve.'

The display was designed to mark the sixtieth anniversary of the establishment of ICI. 'We were going to paint the floor here white,' Bud remembered, 'to denote the fact that you were entering a new and important section. Unfortunately there was a strike and we couldn't get it the way we wanted.'

Dr Bud considered why dyes were so important. 'The textile industry was the big industry, so the moment you've got an artificial dye, you've got something that you can plug into the biggest industry. The other thing is the impact that this all has on people's vision of what industrial chemistry is going to be. Even towards the end of the nineteenth century it is still a shock technology. Together with electricity, synthetic dyes made people think, "What's next?" Drugs followed, then synthetic fibres and plastics, and they are all based on the vision and success of the synthetic colours. Perkin's impact could be quantified in inspirational terms – in what science was now going to do to the world. Perkin's world is one of enormous uncertainty about what the future holds: all you know is that the future is going to be different from the past.

'The really big vision of how chemistry could be applied to industry had come earlier with Davy and Faraday developing new materials and later with Liebig's fertilisers. But dyestuffs were an example of how you apply an entire discipline. It enabled chemists to answer the question "What do you do?" with a practical demonstration.'

Dr Bud returned to his office with the thought that if he were to build the display again he might concentrate more on the romantic side of the Perkin story, 'not just on dyes as a product. But in 1986 we weren't living in a postmodern world.'

9
POISONING THE CLIENTELE

In a continuing effort at precision reporting, a *New York Times* reporter covering the O. J. Simpson trial found himself at a loss in describing the colour of defence lawyer Johnnie Cochran's double-breasted suit. He dispatched a research assistant to a local drugstore to purchase a box of Crayola crayons, from which he selected the closest possible colour: periwinkle. In an informal poll, courthouse wags opted for the less precise purple. During a courtroom break, though, Cochran insisted his suit was blue. 'Just don't call it mauve,' he said.

USA Today, February 1995

In 1870, a German chemist named Dr Springmuhl gathered fourteen samples of commercial magenta dye from his friends in Europe and checked them for traces of arsenic. Some of his findings were alarming. He found that nine of the samples consisted of at least 2 per cent arsenic, and five of them contained between 4.3 and 6.5 per cent. The colour was poisoning its purchasers.

His analysis was sponsored by the German government, some members of which had become concerned by newspaper reports that aniline dyes had caused inflammations on women's skin. Under much pressure from its dye companies, the nation's fastest-growing industry, the German government declared that women had nothing to fear. Dr Springmuhl's work further examined the effect of his dyes on a square foot of wool, and found that while the dye-bath still contained high levels of arsenic poison, only a tiny fraction – 0.0001 of a gram – transferred to the surface of the dyed fabric.

Much to the relief of BASF and AGFA, the government chose to ignore another statistic: while the dried fabric was deemed safe, substantial traces of arsenic were found in water after the first wash, suggesting that the cheaper dyes were far from colour-fast, and presented a considerable risk after exposure to rain or perspiration.

The dilemma was fierce but not new. In 1862, William Cowper, Chief Commissioner of Works, had asked August Hofmann to examine the green compounds in a woman's ballroom head-dress. He found it rich in arsenical Schweinfurt green. Hofmann noted that the same substance was used in many ball gowns, and frequently in wallpaper; it had been banned from such use in Bavaria. He concluded: 'It will, I think, be admitted that the arsenic-crowned Queen of the Ball, whirling along in an arsenic cloud, presents under no circumstances a very attractive object of contemplation; but the spectacle – does it not become truly melancholy when our thoughts turn to the poor poisoned artiste who wove the gay wreath, in the endeavour to prolong a sickly and miserable existence already undermined by this destructive occupation.'

In the mid-1860s, the Swiss dye firm of J. J. Muller-Pack and Co. faced a health crisis extreme enough to lead to its collapse. The company had leased land on two sites from the traditional dyewood firm J. G. Geigy, and produced mauve, fuchsine and aniline violet, blue and green. The director of the company had visited Manchester to secure the rights for aniline black from Roberts, Dale and Co., and he may have met Perkin to discuss other colours. Business boomed until 1864, when people living near its two factories in Basle began falling ill from water drawn from their wells. In one notorious case, the family and staff of a wealthy landowner became sick after drinking tea.

An investigation revealed that local water had a 'strange disgusting smell that was not well defined . . . You could not get this smell from normal drinking water.' Muller-Pack and Co. was found guilty of arsenic pollution, and was fined and forced to close. The founder of the company was humiliated by a court order to hand-deliver clean drinking water to the locals. New anti-pollution legislation had some effect on the behaviour of other aniline companies in Europe, particularly the desire in Germany to move factories to the banks of the Rhine, where effluent would be broken up by the current and dissolved from sight.

But the health debate would not subside. Its next battleground was England, and in particular the pages of *The Times* and the medical journals in 1884. The main issue now was fastness: although many of the earliest dyes retained their intense colours after many washings, less skilled and less principled dyers were now producing products of inferior quality, often with injurious results. A few years after the German enquiry, a letter appeared in *The Times* from a correspondent calling himself Paterfamilias. 'The unusually hot season has given rise to numerous complaints of unaccountable skin eruptions, which have been attributed to the heat or a feverish state of the system. The real cause, however, is often to be found in the exhibit of Mr Startin, MRCS, in the east quadrant of the Health Exhibition, which contains some horrible examples of skin disease, caused by wearing, when in a state of perspiration, hosiery and flannel coloured with aniline dyes.'

James Startin was a surgeon and lecturer at the St John's Hospital for Skin Diseases in London, and had filled a glass display case with models showing the awful effects of aniline on skin. There were photos too – terrible purulent eruptions. Next to these was a collection of natural dyes from vegetable and insects, including indigo, cochineal and safflower, which were perfectly harmless. Paterfamilias observed that indigo was used as much as ever in England, chiefly because no artificial substitute had yet been found. 'But the brilliant and permanent scarlet produced by lac dye or cochineal is seldom to be found now . . . The assertion that aniline dyes are only injurious when they contain arsenic is quite false; they are chiefly noxious because of their volatility.'

Four days later, *The Times* ran another letter from William Gowland of Essex. He wrote of his knowledge of a woman who purchased red silk stockings near Charing Cross, and later found that her skin was in such a high state of inflammation that she felt compelled to consult a doctor. The doctor pronounced the stockings dyed with aniline and therefore 'poisonous'. Mr Gowland trusted that the public would give their full attention to this important matter.

Another correspondent, writing as Anti-Aniline, had also visited, or perhaps owned, a stall in the east quadrant of the International Health Exhibition, and found that it provided relief from the perils of the new dyes. Here you could buy stockings 'made by the peasants of Donegal' coloured entirely with vegetable dyes. 'Notwithstanding all our boasted advance in technical education it would seem that the efforts of our chemists and dyers have resulted in the production of a volatile substitute for the old fast dyes formerly in use, which is calculated to deceive the consumer to the benefit of the manufacturer and shopkeeper, and ultimately to degrade the character of our coloured textile manufactures. The Shah of Persia was more foreseeing than Englishmen, for he at once discovered the fugitive nature of aniline dyes, and excluded their importation lest they should injure the good name of Persian carpets.'

A few days later the news worsened. Wearing coal-tar colours was only a small problem compared to the horrors of eating them. 'There is every reason to fear that in these "cheap and nasty" times aniline dyes are being used to a considerable extent in the manufacture of sweets and confectionery,' wrote Detector from Mincing Lane. Detector had recently seen the analysis of one type of yellow sweet and was shocked to learn it contained 50 per cent picric acid, 'a poison hardly less deadly than arsenic itself; the rest was probably chalk'. Some dyemakers had refused to accept orders from confectioners, but not all. 'The Shah of Persia is not the only authority by whom the use of aniline dyes has been forbidden, for the Swedish government will not, I believe, allow them in the country. Here, in this gloriously free country, all appears to be allowed, even to the slow poisoning of our little ones.'

The final accusatory correspondent – and there were none who wrote in favour of the new dyes – remarked on another detrimental effect, namely 'the altered appearance of most of the carpets and curtains of the present day after a comparatively short exposure to the light, on account of the fugitive nature of the dye used. Aniline colours are largely used by dyers on account of their convenience and the facility of application; but the public suffer immensely, and the remedy is in their hands.'

The *Times* leader writers believed that the remedy lay in their hands as well. It noted that these complaints had been made for a number of years, but there was hope that the extent of the evil had been exaggerated. 'The wearers of aniline dyed fabrics are countless, while the persons who suffer from them are at least exceptional; and, so far, no cases of aniline poisoning by confectionery have come under our notice.'

The editorial was suspicious of the severity of most claims, and supported the relentless industrial process. Short of legislation banning all artificial dyes, and thereby gravely damaging dye works and the textile trade, there was little to be done except advise its readers to be vigilant. The revolution in synthetic colour had already gone too far. 'The suggestion that the use of the dyes should be abandoned in favour of cochineal, indigo, madder and other animal or vegetable substances, is unpractical, because the supply of these substances is limited, and has been far outgrown by the demand for coloured goods. It would now be impossible to return to what we may call the pre-aniline stage of manufacture, and we must be content with the enforcement of such precautions as may banish or minimise the risk of injury.'

Rather than follow Persia or Sweden (whose restrictions were largely dictated by protectionist rather than health concerns), *The Times* trusted to the responsibilities of British traders. 'A purchaser who suffered in health and pocket after wearing a given pair of stockings or gloves would unquestionably have his remedy against the seller. Even if the penalty fell in the first instance upon the retailer, it would ultimately reach the manufacturer, and it would be sufficient to hinder the distribution of imperfectly cleansed fabrics as ordinary articles of commerce.'

The main culprit was arsenic acid, used in the oxidation process of several colours. Arsenic was still in limited use, despite an awareness of its devastating effects. Its use in wallpaper and paint was particularly popular, not least in a pale green shade that had caught on in the mid-1860s. Here, arsenite of copper was not just a constituent of the dye but the dye itself, and became known, after its Swedish inventor, as Scheele's Green (Karl Wilhelm Scheele was one of the greatest experimental chemists of the eighteenth century, responsible for groundbreaking work on oxygen and other gases and acids). At Guy's Hospital in London a surgeon had been presented with many patients suffering from sore eyelids and lips and lung and throat complaints, and he was the first to isolate a universal cause. A cheap and widely used type of wallpaper was decorated in green foliage and flowers, the pattern made up in thick relief of arsenite of copper. Under heat or agitation from brushing or cleaning, particles of dust would slowly poison people in the room.

The newspapers and medical journals carried many reports, and caused considerable panic amongst readers. *The Times* noted, 'It was not very uncommon for children who slept in a bedroom thus papered even to die of arsenical poisoning, the true nature of the malady not being discovered until it was too late.'

A committee comprising chemists, dyers and some medical men was formed by the Society of Arts to consider the danger, and concluded that the best policy would be public education. The message appeared to be simple: don't buy green wallpaper. Other colours were promoted and bought instead, but with wide use of aniline dyes the problem returned. Arsenic acid used in the manufacturing process of other dyes had often not been adequately filtered and washed away.

The *British Medical Journal* noted that the problem largely affected the working class. Coal-tar dyes had brought a greater choice of colours to the masses, yet also gave a new interpretation to Thackeray's definition of the Great Unwashed; adorned in the more fugitive hues of low-quality garments, it soon became clear that colour-fastness was best preserved by avoiding detergents. 'The cheaper magenta and scarlet fabrics are much sought after by the poorer labouring classes for underclothing, stockings and trimmings,' noted the *BMJ*. 'The safest plan in regard to clothing is to discard all transient colours, and to be content with as much show as is wholesome. The risk incurred by wearing aniline or arsenic next to an absorptive skin, or by breathing the atmosphere of a room loaded with particles rubbed off the wallpaper, overbalances any ornamental effects which these pigments can afford.'

The dye trade took a little while to respond to this crisis, but then did so with clinical thoroughness. The Society of Dyers and Colourists decided that what was needed was proof, for the public outcry was based on hearsay and inadequate evidence. While the Society acknowledged that the wallpaper incidents had been regrettable, it claimed that by 1870 great pains had been taken to eliminate the problem. It would not accept that its textile dyes had ever caused the problems described in the press.

Accordingly, it had been in correspondence with James Startin, whose shocking display had been at the root of this predicament. From the start, they tried to paint Startin as a charlatan, a doctor who had no idea of their complex trade. 'We want the truth, and well authenticated cases, and proof that aniline dye has produced the injurious effect alleged,' pronounced one member. 'Fortunately, chemistry is a science which deals only with facts. It does not deal with speculative opinions or sensational exhibits.'

Unfortunately, it now had to: dyers acknowledged the untold damage the exhibition and its publicity could inflict on its industry in the long term; already the *London Commercial Record* had reported that customers were specifically asking for goods not dyed with the new colours.

The dyers argued that it was quite possible that arsenic or sulphuric acid could be introduced into the dyeing method of vegetable dyes, perhaps in the mordanting process. It was quite possible that the material itself might be to blame. It was likely that the materials in question had not been dyed with aniline at all, or at least not British aniline, or at least not aniline that was applied with the help of a skilled chemist.

The Society regretted that Mr Startin was obliged to turn down its offer to attend one of its meetings due to other engagements. He did, however, send some small samples from his exhibit, and these would be examined and a report prepared for a future gathering. In the meantime, the colourists were left to dispute Startin's findings with some inexact science of their own. One member, Charles Rawson, mentioned a little practical experiment he had conducted at home and on the streets. He held up two hanks of woollen yarn, one dyed with cochineal, one with aniline scarlet. He told how some of the yarn had been made into socks, and how for a week he had worn the cochineal sock on his left foot, and the aniline scarlet one on his right. 'My brother has made a similar experiment,' he explained, 'and in neither case has the slightest irritation or inconvenience been experienced.' Mr Rawson suggested that like experiments might be conducted on a more extensive scale.

The Society's honorary secretary G. H. France had done some research of his own, writing to dye works throughout Europe asking whether their workers were getting sick. In the factories, people handled dyes in large quantities every day, often with no protection to their hands: were there any cases of skin disease or rash? BASF replied from Ludwigshafen that although it employed a great many hands making many varied dyes, they had never heard of any complaints. Leopold Cassella and Co., from Frankfurt am Main, said they were producing 30 classes of coal-tar colours, but that arsenic was only used in one of them, magenta. In this case, arsenic made up only 1 per cent of magenta crystals, and 95 per cent of this remained in the dye bath. 'We may also state that the workmen employed in the manufacture have always their skin and saliva intensely dyed with the colour, and, notwithstanding this, the men are perfectly well and healthy, and their mortality compares very favourably with that of other workmen who never come near the dyes.' The German government had appointed an expert committee to conduct their own investigation, and it concluded that the dyes were quite harmless. The committee included August Wilhelm Hofmann. He and his colleagues reasoned that the dyes were so safe they could happily be incorporated into foodstuffs.

In Britain, Ivan Levinstein reported from Blackley, near Manchester. He said that in all his years as a colour maker he could not recall one single case of skin eruption on his staff. Williams Bros and Ekin, from Hounslow, Middlesex, mentioned that an analytical chemist from London called Antony Nesbitt fed his rabbits for many weeks on oats which had been steeped in strong solutions of magenta, violet, brown and orange. The rabbits seemed to like it, and stayed white.

Three weeks later, the Society met again to hear the results of its analysis of James Startin's exhibits. Two chemists had examined eight tiny samples, each about one-inch square. These were cut from offending flannels, socks, stockings and gloves, and dyed with both aniline and vegetable colours – logwood, magenta, alizarin and safranine. The results, alas, were inconclusive. No arsenic was detected on the samples. The assembled dyers reasoned that some of the dyes were of a cheap and primitive form, and unworthy of their profession. Others were more fugitive than those being produced currently. In all cases there was no proven link between dyes and skin disorders.

William Perkin, a founding member of the Society of Dyers and Colourists and later its president, probably observed this investigation with some satisfaction and a little bemusement. It was not the first time his discovery had been the subject of public scrutiny; in earlier days he had been accused of polluting the canal and local streams, and destroying ancient trades and the lives of those who lived off madder. He would be accused of making colours more fugitive than those they replaced, and of endangering the lives of his workers and those living near his works.

At the time of the latest skin scandal he was probably relieved that he was no longer part of the industry.

In 1873 he had sold Perkin and Sons to Brooke, Simpson and Spiller, the successors to Simpson, Maule and Nicholson. He was thirty-five, and it had been seventeen years since he had made mauve. He had several reasons for the sale: his firm was too small to compete, for he had fallen victim to his own success with alizarin; any expansion would have required employing more chemists, and he believed that there were too few qualified English scientists and he would have a tough time recruiting from Germany; in the last year there had been two bad accidents on his site following leakages in his retorts, and as a deeply religious man he did not feel he could continue to expose his staff to danger. In addition, the price of natural madder had recently fallen by almost a third in order to compete with alizarin, and thus Perkin had been forced to reduce his own price and profits accordingly; he felt inadequately protected by his patents, and disliked spending time in litigation. Another reason came from abroad: whenever Perkin looked towards Germany, where many of his former trainees now worked, he saw only mass investment and expansion, and rigorous state protection. He saw no prospect of a happy future.

Several years later, after correspondence with Perkin, Heinrich Caro found time at BASF to write an account of Perkin's decision to pull out of the industry he helped create. 'There was no possibility of remaining in the same condition and to rely on the protection of the British patent law to ward off the invasion of German alizarin made at Hochst and Elberfield in infringement of the joint patents of Perkin and the Badische.' Cost was a problem, but so was quality. According to Caro, Perkin was 'fatally prejudiced' towards a process of producing alizarin by a method that had now been improved upon in Germany. The Perkins realised they would have to triple the size of their plant, and probably relocate from Greenford Green. So they sold up, the 'prosperity of alizarin being past and gone'.

The sale of his works was an eventful procedure, but seemed to turn out satisfactorily, at least until a writ arrived. Initially, the Perkins offered their factory to BASF, but its directors declined. Then Thomas Perkin met a rival dyer Edward Brooke in a train carriage, and they began talking about the future. Brooke said that his firm, Brooke, Simpson and Spiller of Hackney Wick, was keen to make alizarin, and asked whether the Perkins would consider some form of partnership. This was rejected, but a little later the possibility of an outright sale was mentioned.

The Perkins suggested a sum of £110,000, not including stocks, supplies and future orders valued in the region of £35,000–£50,000 for making the red shade and blue shade of alizarin, and also mauve, violets numbers 1 and 2, magenta, and black powder. The ingredients in stock included alumina, petroleum, acetic acid, alum cake, aniline, bromine, china clay, methylated spirit, coal, manganese, caustic soda, muriatic acid, naphtha, nitric acid, foaming Nordhausen acid, sulphuric acid, sulphate of potash, soda crystals and chloride of lime. Edward Brooke was very interested. He

William Perkin in black-and-white: a self-portrait at 14; Dr August Hofmann, who ıought his protégé was wasting his time; Perkin the inventor with his second wife lexandrine in 1870.

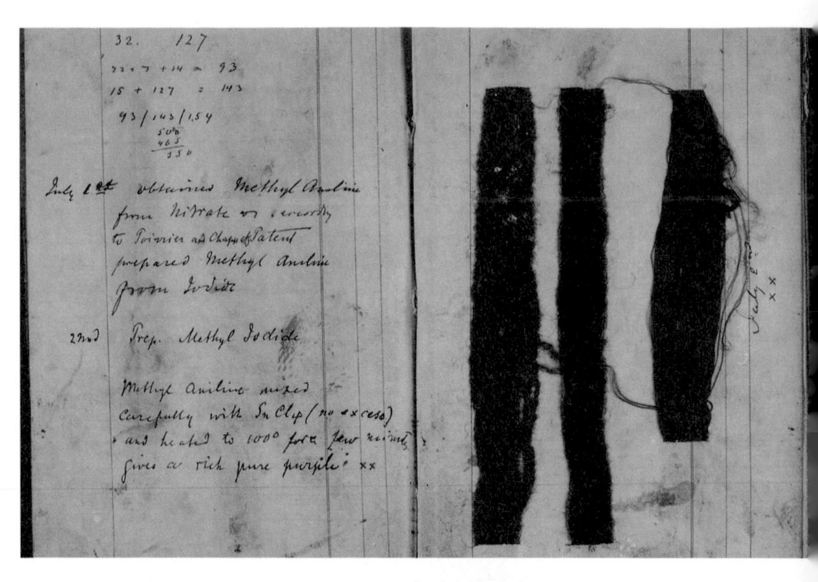

32. 127

15 + 127 = 143

93 / 143 / 154

July 1st obtained Methyl Aniline from Nitrate

to Poirrier and Chappat's Patent prepared Methyl Aniline from Iodide

2nd Prep. Methyl Iodide

Methyl aniline mixed carefully with SnCl4 (no excess) and heated to 100° for a few minutes gives a rich pure purple xx

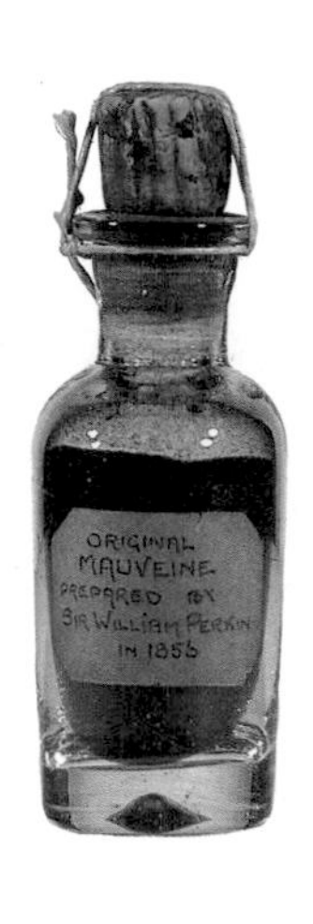

2 'The torch which enlightens the path of the explorer in the dark regions of the interi or of the molecule' — but to the dyer all that mattered was brilliance: a recipe book fro 1868, and the mauve, alizarin, crystals and cloth that changed the colour of our streets.

'Luring on foolish bachelors to sudden proposals': a silk dress dyed with original
.uve in 1862.

4 An object of ridicule, an object of pride: gaudy, fugitive and poisonous, but the new colours proved irresistible to fashionable Paris and London.

A miracle that Perkin did not blow himself and Greenford Green to pieces': Perkin
cond from right), his brother Thomas (second from left) and fellow dyemakers
ight in their continued good health in the field outside their dye works. The sketches
ɔw the expansion from 1858 to 1873.

6 Fame at last: at the British Association meeting in 1906, Perkin's illustrious colleagues paid tribute to a pioneer, but his name would soon be forgotten.

he grand old master at 68, and the house that mauve built. The Perkin Medal has
e been won for innovations in nylon, vitamins and medicines.

8 What dyes did next: a catwalk model in Paco Rabanne, and a stained micrograph the serpentine cords of tuberculosis bacteria.

believed the matter had to be conducted in secrecy, lest an interest from other companies drove up the price. His initial correspondence referred to 'a certain matter ventilated a few days ago' and called for a formal meeting at his office. 'We will be ready,' he wrote.

Several meetings followed in November 1873, including an inspection of the books at the Perkin factory and a tour around the works. Thomas Perkin told Brooke that Perkin and Sons had already contracted to sell all the artificial alizarin they could make for the following year, almost 400 tons at a price of 2s 3d per pound. The Perkins calculated that the actual cost of producing alizarin was 1s 6d, and accordingly estimated a profit for 1874 of £30,000. 'I really thought there was now scarcely any limit to its consumption,' Thomas Perkin explained, 'and that consequently new works would have to be provided if the demand was to be met.' He reminded them that at that point synthetic alizarin was being produced nowhere else in England. Accordingly, their site was subject to strict security measures, but Thomas Perkin regarded Edward Brooke as a highly trustworthy businessman, albeit a competitor, and he 'felt sure that he would not take advantage of anything he saw even if the proposed purchase fell through'.

Before the sale was completed, the Perkins felt obliged to describe a few handicaps particular to the manufacture of their successful colours. The Greenford Green site had two great defects – an inadequate water supply and the entire absence of suitable drainage. The site sloped from north to south and was heavily waterlogged from the canal. Multicoloured waste regularly found its way back into the canal, the nearby brook and the River Brent.

Alizarin involved a far more complex and laborious process than mauve or violet or magenta. 'I warned them that the purchase of the crude anthracene was really as important as the manufacture of the alizarin from it, and required great care and skill in the selection,' Thomas Perkin remembered. In addition, manufacture was 'very heavy', and required one of their chemists to be in attendance at all times. 'We attributed a good deal of our success to my residing very near the works so as to be able to be constantly there whenever anything important was going on . . . when I was away from the works for only one week the production fell off, and during 1873 when I was on the continent for three weeks the production fell off to the amount of several thousands of pounds.'

Despite these details, Brooke, Simpson and Spiller swiftly agreed to a purchase fee of £105,000, including all patents, at the end of December 1873. William Perkin recorded how the final figure was settled over wine and biscuits, and that he and his brother agreed to be available for consultation and assistance for the first six months after the sale. The first act of the new owners was to stamp over the Perkins' invoice books with their own name. They then hired two Perkin and Sons' technicians, Mr Brown and Mr Stocks.

Eight months later, the works at Greenford Green were almost at a standstill. The client list was destroyed, the entire production of colour ravaged by mismanagement and pigheadedness. The damage was laid out in a writ filed by Brooke, Simpson and Spiller in September 1874. The new owners claimed they were not to blame, but instead accused William and Thomas Perkin of deception bordering on fraud.

The new owners claimed they had been entirely misled as to the true value of the business and its profits. They said there had been an inadequate system of book-keeping, and that they were never shown the full accounts when they asked for them. They claimed that Thomas Perkin had assured them that they would get the full value of their investment back within two or three years, but this now seemed an impossibility. 'Now that we are able to go into facts and figures for ourselves,' Edward Brooke wrote to Thomas Perkin's solicitor, 'we are alarmed to find that we have been fearfully, but we are sure unintentionally, deceived by your clients in the cost of the manufactured product.'

No matter what they did, the new owners were unable to make a profit. Brooke said he was told that the cost of making alizarin was between 1s 4d and 1s 6d per pound, but found that the true cost was 1s 11d per pound, to be sold under contract at 2s 3d per pound with a £6 discount per 100 pounds. Consequently there seemed no possibility of achieving profits on the existing contracts. Brooke Simpson and Spiller thus demanded their money back, plus an additional £5,000 'to compensate us for our loss, trouble and disappointment'.

The Perkins declined this offer. They were hit with more complaints. After another estimate a few weeks later, Edward Brooke discovered that the actual cost of manufacturing alizarin had slightly exceeded 2s 3d per pound, and that the £6 discount meant they faced losses of more than £3,000. He also claimed that when his colleague Richard Simpson had visited the works he had noticed 25 retorts set in brickwork and carefully whitewashed. He was told that these were superfluous to their present needs, but could be used as replacements if required. Simpson had since discovered that they were all cracked and worthless.

William and Thomas Perkin prepared a stinging rebuttal. Above all, they were wounded that their reputation should be threatened in this way, and that such allegations were now placed on file at the Public Record Office. Thomas Perkin believed that the detail of Edward Brooke's recollection of his visits to the site and their negotiations were 'totally imaginary'. He claimed that he was given full and frank access to the books, and that 'no conversation whatever took place [regarding] the cost of manufacture of alizarin or the quantity of alizarin that could be manufactured or the profits that might be realised from it'.

Thomas Perkin then ridiculed the method by which Brooke, Simpson and Spiller had tried to make alizarin, claiming that any attempt by the Perkins to offer advice was dismissed as unnecessary interference. Looking at Brooke's method of making 34 tons of alizarin, Perkin noted with alarm how 'the quantity of crude anthracene exceeds the quantity required by 16 tons . . . the amount of light oils is not sufficient by 120 gallons . . . the chloride of lime is in excess by 4 tons . . . the sulphuric acid is in excess by 6 tons . . .'

As to the retorts, Perkin rejected the insinuation that they were whitewashed to deceive the plaintiffs. They were out of use due to their foul stench, and were currently being cleaned. Only one was cracked, the remainder being used for the first three months after Brook, Simpson and Spiller had bought the works.

The new owners also mishandled the disposal of toxic waste, releasing calcium chloride into the Grand Junction Canal, thereby visibly polluting the water and ground for a quarter of a mile surrounding the works. Worse still, they failed to maintain the existing water pumps and thus drew contaminated water from the canal back into the production process.

The Perkins' conclusion was harsh. Since their purchase of the works, the brothers believed, Brooke and partners had managed the business carelessly and injudiciously. 'Instead of looking after it themselves and personally superintending the working of the various processes as we had done, they seldom went over the works or remained on the premises more than two or three hours at one time . . . they also very largely increased many of the expenses and outgoings of the business in a way which in our judgement was utterly unnecessary.' Their insurance and rates bill, for instance, had gone up from the Perkins' annual payment of £228 to £1,594 in just the first six months, suggesting colossal financial ineptitude.

In March 1875, a judge ruled that the plaintiffs did not have a sufficient case, and thus threw out their claims, awarding costs to the Perkins. Subsequently they reflected on why Brooke, Simpson and Spiller, previously a highly reputable company, might have chosen to act in this way, and manage their business so poorly. 'We are reluctantly compelled to believe . . . that their continued and persistent opposition to all arguments and advice was wilful and deliberate, and intended to produce loss and damage the business.' The Perkins hinted at a conspiracy, perhaps believing that Brooke, Simpson and Spiller were exacting a painful revenge. Not only were the company's founders envious of Perkin and Sons' success with alizarin, but they were still resentful of an earlier decision by the Perkins not to purchase raw materials from them (when Brooke, Simpson and Spiller took over the firm of Simpson, Maule and Nicholson in 1868, the latter regularly conducted business with William Perkin, selling him aniline and nitrobenzene). Their claim was thus intended to ruin the Perkins financially and by reputation. If Brooke, Simpson and Spiller had succeeded, they would have caused enormous damage to the Perkins' trade and rendered the future sale of their business extremely difficult.

As it was, it was Brooke, Simpson and Spiller who faced the future with apprehension. Within a year of the litigation, they tried to sell the works on. The firm continued to lose money for the next eighteen months, and destroyed all of the advantages established by the Perkins. It tried many ways to stem its losses. One of its first decisions was to cease immediately the production of mauve.

Part Two
EXPLOITATION

10
RED LETTER DAYS

The tyranny of mauve is over! That's the word from Vivian Kistler, a member of the Color Marketing Group, an organization that helps decide which colors we wear and decorate our homes in. There's no doubt that mauve is now dead as a fashion color, Kistler said. Even firms that manufacture rubber and plastic kitchen products such as dish drainers are phasing out their mauve lines. 'In the last six or seven years we have been mauved to death,' Kistler said, laughing.

The *St Louis Post-Dispatch* at the 12th annual New York ArtExpo art dealers' trade show, 1990

The fashion caravan sashayed into Milan yesterday for the third leg of its month-long tour, due to end in Paris in a fortnight. Milan is the city where fashion's most influential labels, such as Gucci, Prada, Versace and Dolce & Gabbana, show their collections and set the season's trends, which are quickly appropriated en masse by the high street. There were plenty of modern and supremely wearable pieces for those who like a bit of Versace-style flash glamour, from the mauve cropped sheepskin jacket and the lady-chic purple trouser suit to the claret velvet tailored jacket and the streamlined camel car coat.

But the Italian fashion parade says mauve is back, *Independent*, February 2000

A wealthy man at thirty-six, Perkin built his dream home and called it The Chestnuts. His previous house in Sudbury he fitted out as his new laboratory. He found more time for music, and for the committees of chemical societies and church councils. He spent his money – a sum approaching £100,000 – on charity and local property. Sometimes he played woodwind instruments.

And in this way one of the greatest chemists of his age opted out. Unattached to either academic institution or industrial concern, he chose to continue his researches in seclusion, contributing the occasional piece of important work to the journals but never again making such a significant contribution to the trade of his country. His reputation was secure, although he knew that some regarded his talents as ill-spent. The questions were still asked: What use was a colour? How much was a colour worth to science? Did Perkin really want to be remembered for turning the streets purple?

Perkin himself appeared content. He had proved his doubters wrong. He had made a lot of money, and the view from his garden stretched for a stunning green acre. He had created radiance out of basest residue, and in so doing provided the key to other people's goldmines. But perhaps there was anxiety too: he had contributed to carbon chemistry, but he had not fulfilled his original quest, the synthesis of quinine, an achievement that would have improved the health of millions. He was certainly dismayed at the exploitation of his invention overseas as the British dye trade lost ground. And perhaps he saw his work having little further benefit in the years to come. He would have been surprised.

'My personal memories of him go back to about the year 1880, when I was six,' his nephew Arthur Waters recalled. 'Uncle William's visits were always red-letter days with us, and I have vivid memories of a wonderful show of conjuring tricks.' Perkin liked to get to Waters' house on his bicycle, and usually arrived with his black clothes covered in chalk dust. Coal-tar had not yet covered the roads.

A local writer from Harrow recalled 'happy summer days in the field now Sudbury Recreation Ground, where William was wont to gather the youngsters around him by sounding a trombone and quietly enjoying the spectacle of seeing them scramble after the sweets and other delicacies which he scattered among them'.

His brother Thomas became known locally as the Squire of Greenford, spending much time on horseback around his farm and on the Greenford Drag Hunt. He became a churchwarden, and continued to play several stringed instruments, including a Stradivarius violin.

William Perkin's first wife Jemima had died in 1862, and four years later he married a Polish woman called Alexandrine Caroline Mollwo, who bore him three daughters and a son. 'Aunt Sasha (as we called her) was a good housekeeper, and liked everything well ordered,' Waters remembered. 'The house was beautifully furnished in the mid-Victorian style, and there was a large and lovely garden, and hothouses produced the most delicious grapes I have ever tasted. Uncle William was a vegetarian, and certainly knew how to grow the best.'

Perkin became an evangelical churchman, organising weekly meetings with visiting preachers and raising money for a new barrel-organ for the hymns. Personally, he preached charity, moderation and abstinence from alcohol, and sometimes these beliefs combined. 'I always felt an interest in this neighbourhood,' he told the *Harrow Observer*. 'I and some of my neighbours thought that something might be done for the big lads and working men of Sudbury, and we hired some of the cottages in the paddocks (now the Sunday school) and also a shed opposite which had been used for for carriages in connection with the racecourse at the back, which fortunately for the neighbourhood was not a success. These cottages were used for a working men's club and institute, and the shed for lectures and amateur concerts.'

The club was only a short-lived venture, because the working men did like a drink. 'It was difficult to keep in hand on account of the lively spirits connected with it, and I felt rather unhappy on account of its purely secular character.'

At home Perkin read the journals, in which he followed Germany's dominance of the industry he founded. He read that Pullars of Perth had moved into the revolutionary art of dry cleaning. In 1875 he probably saw an account of how Brooke, Simpson and Spiller had sold his old factory to Burt, Boulton and Heywood. By then, most of his customers had been driven to buy alizarin abroad. Burt, Boulton and Heywood immediately faced claims that the old Perkin factory had caused irreversible environmental damage, including water pollution, and so it closed the factory down, transferring the works to Silvertown, Essex, a plant which later become part of the British Alizarine Company.

Elsewhere, there was the prospect that Perkin's dyes would enjoy artistic posterity. The textile designer William Morris experimented with synthetically dyed wool and silk in some of his earliest embroideries (although he later suspected their fastness and rejected them in favour of natural recipes). Perkin's dyes were now also being used as artist's pigments, and these aniline insoluble 'lakes' had given rise to the new shades of vermilionette, Post Office red, and emerald tint green. The *Journal of the Society of Dyers and Colourists* noted that the government was using the reds extensively in its official publications, and that the Manchester Tramway Company had used vermilionette on its cars.

In the galleries, there was disquiet over how swiftly the new colours faded without varnish. In 1877, the science journal *Nature* noted the concerns of Mr Joseph Sidebotham, a member of the Manchester Literary and Philosophical Society, who saw that aniline colours were being used increasingly for tinting photographs. 'Anyone who knows the speedy alteration by light of nearly all of these colours will protest against their use,' *Nature* reported. 'A statement of this with the authority of some of our chemists would probably have the effect of causing them to be discontinued by all artists who care to think that their works should last more than a single year.'

Some years later, the earliest attempts in New York to regulate the contents of food caused a scare over the use of aniline dyes as a colorant. They were used extensively in sausages and jams and confectionery and baking. Perkin himself would be drawn in, as an American reporter accused him of poisoning his people. 'I would not like to take sides in that matter,' he said. 'It is probable that there have been abuses of the uses of aniline dyes in foodstuffs. In fact, I know there have been. But this is certain: the amount of aniline dye necessary to colour a food is so minute that if the same quantity of strychnine were used it would be equally harmless.'

Perkin's colours travelled the world on postage stamps. When, in the mid-1860s, the American Civil War caused a severe cotton famine, dye firms were forced to look for new markets. In Lancashire, the firm of Roberts, Dale & Co. had a significant breakthrough when its chemist Heinrich Caro struck up a relationship with the London banknote and stamp-printing firm Thomas De la Rue.

In 1863, Hugo Muller at De la Rue expressed an interest in using mauve colour lakes as a replacement for its unreliable cochineal dyes. Muller had worked with Justus von Liebig in Germany, and had been recommended to De la Rue by August Hofmann. It is unknown why Muller chose to use Caro rather than Perkin for his supply of mauve (perhaps it was merely due to their shared language), but he was clearly pleased with the results. 'These colours are so magnificent that I would make every effort to use them,' Muller wrote to Caro in July 1863. Within a year, several aniline colours, including Hofmann's violets and aniline black, were used to colour half-penny, penny and sixpenny stamps. It is possible that an aniline dye was used for later printings of the Penny Black, and Perkin was proud to note in a speech of 1887 that his mauve had been used since 1881 to send messages and love poems around the world. The stamps were used for the remainder of Queen Victoria's reign, and were not withdrawn until 31

December 1901. But this is likely to have been the original mauve's last commercial use.

Freed from the demands of industry, Perkin renewed his acquaintance with pure laboratory research, and made significant advances. Between 1874 and his death in 1907 he published over sixty scientific papers, most concerned with magnetic rotary power and the molecular architecture of a long list of increasingly useful chemical compounds. He worked on low-temperature combustion, revisited the constitution of his earliest dyes, and progressed from colour to scents. His nephew Arthur Waters remembered the smell of violets coming from his laboratory, although Perkin first produced synthetic coumarin, a scent associated with new-mown hay and the tonka bean.

Throughout his industrial career, one colour eluded him and his colleagues, and remained the Holy Grail of the synthetic dye trade until the end of the century: indigo. It was perhaps inevitable that its synthesis would first emerge in Germany, and predictable that its formula was made possible by a specific chemical reaction with its roots in Sudbury.

Natural indigo derived from the plant *Indigofera tinctoria*, and came predominantly from India, where its preparation and use were first described to European traders by Marco Polo. It arrived in Verona and London in the thirteenth century, but was more expensive than woad and was used chiefly as an artist's pigment. But it was in the sixteenth century, with the Dutch and British opening up of the East India trade, that the dye competed with spices for space on the shipping routes.

Several European countries banned 'the devil's colour' on account of its effect on their native woad production, but England actively welcomed the trade from its own colonies, and the colour became increasingly fashionable as a textile dye in Paris and London. By 1770 Great Britain was importing almost one million pounds a year from its plantations in Bengal and South Carolina. A century later, India possessed 2,800 indigo factories, making blues for sailors' uniforms, the Union flag and the army's woollen greatcoats. None of them could have foreseen the synthetic revolution to come.

In Berlin and Munich a pupil of Robert Bunsen named Adolf von Baeyer had been working on the formula of artificial indigo since 1865. His progress was aided by William Perkin's synthesis of cinnamic acid from benzaldehyde, and in 1880 he succeeded in preparing the dye in test-tube quantities. Money poured in from Hoechst and BASF (money obtained from the success of alizarin), but industrial production was hampered by the high cost of the base material toluene.*

It is possible that Perkin first heard about Baeyer's advances when Heinrich Caro wrote to him from Mannheim on 10 December 1881. The letter reached him at Smedley's Hydropathic Establishment, Matlock Bridge, Derbyshire.

My Dear Mr Perkin,

Let me now first of all hope that your present stay at Matlock may not be caused by anything further than your confidence in the restorative effects of hydropathic treatment upon the general system, and that you may now feel again refreshed and rested. [There is no record of Perkin being unwell]. You have gone through an astonishing amount of real hard work . . .

Baeyer has just been deeply afflicted by a terrible family bereavement. Perhaps you have heard that he lost his eldest son, a fine boy of eleven years by [the bite of] a mad dog. The accident happened in Flims, Switzerland, in August last, and the fatal consequence has followed quickly afterwards. Baeyer has now recovered from the terrible shock and is more active than ever . . . The theorising part of his investigations in the indigo group may now be considered, and you may soon read his elaborate work.

* Heinrich Caro worked with Baeyer on the indigo process, and Caro's patent described the modification of the Perkin process. In practical use, the dye was produced on the fabric during the last stage of printing.

In 1890, Karl Heumann, a researcher at the Swiss Federal Polytechnic in Zurich, discovered a new route to indigo using aniline and naphthalene, a synthesis using a new arrangement of the carbons that produced mauve and the very first coal-tar colours. This new process enabled BASF to control the supply of synthetic indigo, and within two decades it had devastated the traditional supplies from India.

While William Perkin took great pride in the chemical achievement, he received only a tiny fraction of the credit for its inception, and none of the fortune that accrued upon its production. The most sought-after colour had its roots in his factory, but was now being controlled abroad. He read in the *Journal of the Society of Dyers and Colourists* that BASF had spent hundreds of thousands of pounds on the commercial production of indigo in the 1890s, an investment that would yield a staggering 152 German patents.

Perkin discussed the synthesis of indigo with his sons, who all studied chemistry and enjoyed distinguished careers. Like their father, they attended City of London School, and worked in a makeshift home laboratory during the holidays. Perkin's eldest son, William Henry Perkin Jnr, would later work in Munich with Baeyer, and specialised in work on alkaloids such as strychnine and the synthesis of substances whose molecules contained carbon rings (before Perkin Jnr, it was believed that no ring with less than six carbon atoms could exist, but in 1884 he prepared rings with four). Perkin's son had strong associations with Oxford University, where he was elected Waynflete Professor of Chemistry and Fellow of Magdalen College, but he also studied in Manchester, where he was nicknamed W.G. on account of the resemblance of both his beard and his bowling action to W. G. Grace.

He claimed as his assistant Chaim Weizmann, later to become the first president of the State of Israel. Weizmann worked as a dye chemist at the Clayton Aniline Company, where he also developed a commercial synthesis of camphor, used in medicine as a liniment.

The Clayton Aniline Company had been founded by Charles Dreyfus, a Zionist and campaign manager for prime minister A. J. Balfour; it was Dreyfus who introduced Weizmann to Balfour in 1906, an alliance that led eventually to the declaration that pledged British support for a Jewish homeland in Palestine.

Within industry, W. H. Perkin Jnr's greatest popular achievement concerned non-inflammable underwear. His manufacturing experience led to consultancy work for large companies, and he was asked to solve the dilemma of flannelette – a cotton with raised fibres and a wool-like touch – that was frequently used among the poor as cheap undergarments. Flannelette tended to 'flash' when exposed to extreme heat, and caused several deaths, especially among children. Perkin devised a fire-proof coating.

Arthur George Perkin followed his father to the Royal College of Chemistry in 1878, which by then had moved to South Kensington. His career was spent with dyestuffs, mostly blue: he was particularly concerned with the chemical properties of natural colours, but he also became manager of a Manchester firm making artificial alizarin. In 1905 he was engaged by the Indian government to supervise a study at Leeds University concerned with scientific ways to improve the production and prospects of natural indigo through improved fertilisers and better methods of extraction. But he was forced to concede that small improvements were ultimately futile in the face of the cheaper and more reliable artificial product from Germany.

In north London, his father spent much of his time picking up titles and science medals. In 1883 he became president of the Chemical Society, and a year later he was president of the Society of Chemical Industry. He was vice-president of the Royal Society from 1893 to 1894, master of the Leathersellers Company in 1896, and president of the Society of Dyers and Colourists and of the Faraday Society in 1907. The big societies awarded him the Davy Medal, the Longstaff Medal and the Albert Medal.

Despite this acclaim, few outside the chemical world knew who he was. His friends in the church knew only of his most basic achievements, and he was seldom bothered by the press. Perkin liked it like this. But things changed in 1906, the jubilee of the discovery of mauve.

The year began with King Edward offering him a knighthood, and some reluctance to accept. His sons persuaded him that he owed this honour to himself, and to them, and to chemistry.

'Dear Papa,' W. H. Perkin Jnr began a letter in February 1906,

> We are all, of course, delighted at the prospect of your getting real recognition for your work at last and especially in consideration of the fact that you belong to that rather small body of workers who have never strived to advertise themselves in any way . . . Any honour which is conferred on you . . . is also an honour done to science and especially to industry, and you could not decline without doing an injustice and causing great disappointment. As you know, there has long been a feeling in scientific circles that research work has never been adequately recognised by our past Governments, as it is in Germany, and this is therefore a welcome step . . . *

A few weeks later, Perkin became a celebrity, a forgotten man rescued by an anniversary. After a while, the attention seemed to appeal to him. An international committee met to raise subscriptions for a bust and portrait, and to plan a month of celebrations. Every member of the Chemical Society received a form through the post, which they were asked to send back with a cheque or Post Office Order. The committee was chaired by the society's president Professor Raphael Meldola, once a chemist at Brooke, Simpson and Spiller, who believed that Perkin's reputation was assured, but vague. A year later he observed how, 'with the passing of the

* On 10 March 1906, William Perkin Jnr wrote again, enquiring about the atomic structures of various synthetic perfumes and wishing his father a happy birthday. 'These days a man is still young at 68. A short time ago I bought a very convenient and accurate weighing machine which I keep in the bedroom, and if you have not got one I will send you one as a present.'

generation which witnessed the interest aroused by the discovery of mauve, and which was fanned into temporary excitement by the sensational accounts circulated by the newspapers of the period, the memory of Perkin has faded from the public mind. To most of his fellow countrymen the memorable international gathering in London in 1906 came as a revelation that they could claim as their compatriot the man whom all the nations had sent their representatives to honour . . . *

The celebration was held at the Royal Institution in July 1906, in the very room in which Faraday had lectured to Sir William as a boy. The official programme announced that 'All further information can be obtained from the Stewards (Mauve badges)'. Dinner was taken at the Hotel Metropole in Northumberland Avenue. The following day, Perkin entertained friends at a garden party at his house, and a special train was organised to take his guests from Paddington via Greenford so that they might view his old works at Greenford Green, which were by then tenantless and disintegrating.

During these events, people wore a lot of mauve. As at Delmonico's in New York a few weeks later, his colleagues tried to evaluate all he had done. There was a Perkin for everyone, and for every trade. Raphael Meldola was one of those who took the trip past his factory on the train and seemed disappointed with the impression it made. 'These works would now appear quite insignificant in comparison with one of the great German establishments, and the whole output of dyes during the seventeen years that Perkin was connected with them was not very great as measured by modern standards. Nevertheless it may fairly be said that no single factory established in this country has ever given rise to such world-wide developments.'

* On 28 June 1906, Caro wrote that he was looking forward to coming over to present Perkin with his personal congratulations. 'As you are probably informed, much interest is taken in your jubilee by my German fellow chemists, but I have noticed that the younger generation, being intensely bent upon the refinement of their present and future tasks, has almost lost the "historical sense" availing in our youngest days.'

Hundreds of chemists showed up for the celebrations, some with colour still on their hands. Without these people, far fewer shades of reds and purples and yellows and greens and blues: if they were not dead or dying they were there making speeches in front of Faraday's original sample of benzene, and most of them also took pleasure in holding up a little glass phial containing a darkening lump of original Perkin mauve. They had come from Germany, Switzerland, Italy, Japan, America, Australia, Holland, France and Austria – Liebermann, Ulrich, Baekeland, Caro, Müller, Duisberg and 200 others, all with special stories and promises not to detain their audience unduly.

Carl Duisberg, director of Bayer in Elberfeld, seemed to spend the morning in a dream. He recalled how Faraday's sample of benzene 'expanded to an enormous vessel filled with millions and millions of gallons of this product. I saw all those gigantic factories in which benzene is employed and applied to manifold wonderful purposes . . . There appeared to me the thousands and thousands of coal-tar colours, commencing with mauve and magenta, passing on to artificial alizarin through the large series of rosanilines and azo-colours to the king of all artificial dyes, the synthetical indigo. Not alone these, but also the pharmaceutical industry, with the numerous pharmacological curatives, stood out distinctly before my mind . . .' Before he woke up, Duisberg mentioned carbolic acid, antipyretics, antineuralgics, astringents and hypnotics. He mentioned artificial vanillin and violet and photochemical products, 'and in the midst of this vision there stood out in actuality the man in mental and physical vigour who had founded all this'.

Perkin needed all vigours that Thursday morning, as he rose no less than nine times to pay thanks to the tribes of chemists who had paid thanks to him. He would return to his house that night laden with medals and framed inscriptions on parchment. La Société Chimique de Paris gave him the Lavoisier Medal; the Deutsche Chemische Gesellschaft presented him with the Hofmann Medal;

he got a medal from La Société Industrielle de Mulhouse; the Technische Hochschule of Munich conferred an honorary doctorate. A woman called Nora Hastings read a poem:

> A crown of Fame! Fulfilment of thy work well done,
> And knowledge of people's gratefulness;
> The promise of life's purpose, fully wrought and won,
> And glorified by its great usefulness.

Sir Thomas Wardle, representing the Society of Dyers and Colourists, had come from Bradford to say that he could remember one of the very earliest batches of mauve, for he was one of the first to use it commercially in a dyehouse. He retold a true story that August Hofmann had told him not long before he died. He said that Hofmann was back teaching in Berlin, but had taken some of his students on a trip to America. On one expedition in the north-west they had come across a group of North American Indians, all of them dyed from head to foot in what looked like Perkin's mauve.*

The colour had also reached Japan. Jintaro Takayama, president of the Tokyo Chemical Society, expressed great thanks for mauve, and wished Perkin long health and happiness and further researches advanced with an unimpaired activity. This wish was soon a theme, as all speakers blessed the 68-year-old man with great resilience. Even those who couldn't attend had expressed this hope in Reuters telegrams, and they were read out to big applause. The Sydney section of the Society of Chemical Industry tendered cordial congratulations and hoped that he might long enjoy the pleas

* August Hofmann died in 1892. His legacy was one of limited chemical innovation but great inspiration. Many of his pupils in London and Berlin scaled greater heights of practical achievement, but all acknowledged his early encouragements regarding benzene, toluene and their amino compounds, and his contributions to the theories of molecular formulae. Many reactions still bear his name. In Germany he helped establish the strong patent system for the chemical industry, and was made a baron four years before he died. He married four times, believed to be a record in the field of German chemistry at that time.

ure of labouring for the advancement of chemical knowledge. One telegram was from Dr Bottiger, the man famous for Congo Red, the first cotton dye not to require a mordant for application. He wished Perkin a long life.

Raphael Meldola, the chairman of the celebrations, offered 'the hearty wish that he may yet be spared for many years'. He also observed, with some regret, that 'coal-tar is not a subject which lends itself readily to that hilarity which our visitors expect to be associated with an after-dinner speech'. He believed that the 'one great joke' which centres around the coal-tar industry had been so frequently repeated that it could no longer claim to be funny.

He tried it anyway. 'It may be well to repeat once again that Sir William Perkin did not discover the coal-tar colours by observing the iridescent film on tar distributing itself over the surface of a pool of water.' There was haughty laughter in the audience: 'Some people can be so stupid.' Meldola said there were other misapprehensions about Perkin's colours that had taken much time to correct. He believed that the coal-tar industry 'had had to pass through the phase of actual prejudice. I remember even, in my younger days, the term aniline dye was a term of reproach. A coal-tar dye was looked upon as gaudy, fugitive, and having every objectionable quality.' He said it took a long while for the public to recognise 'what a miserable colourless world this would be' without them.

Sir Robert Pullar, for sixty years a practical dyer, said that it would take a whole evening to speak of the astonishing changes that had occurred in his industry since a man of eighteen had sent him a narrow sample of cloth. But he felt sad about how they were now enthralled by the chemists and techniques of Germany. Germany now made all the best colours. 'I am afraid we shall need to go there for a long time yet to buy them.'

Many guests spoke in German or Dutch, but Professor Duisberg did his best torturing analogies in English. He and 3,500 German chemists were now gardeners in the extensive grounds laid out by

Perkin fifty years ago, and he took unfortunate pleasure in stretching such a natural image for such an artificial breakthrough. 'It has fallen to our lot to assist in cultivating and grafting the young plant planted by him when he invented mauve, and to gather the fruits from the large orchard, full of strong and mighty trees which have grown up to full maturity . . .'

Adolf von Baeyer was only present in spirit, but one of his recent lectures was presented to Perkin bound in mauve-dyed leather, and read. Baeyer had told his students that the key to aniline shades lay in the basic properties of the carbon atom. The rays of aniline colours 'are the torch which enlightens the path of the explorer in the dark regions of the interior of the molecule, and the man who has lit the torch is . . .' just getting up to speak for the sixth time that day.

Perkin was astonished that the mauve had led to this. 'When this year opened, I received a New Year's card from my old friend Hofrath Dr Caro, in which he referred to this year being the golden jubilee of the industry. I little thought that I should hear any more about the matter . . .'

Chemists cheered and laughed and clapped, but the tone of his speech was serious. Perkin wished to remember those who were no longer around, especially his father and brother. He had kind things to say about the French, who had greatly encouraged him by presenting him with his first Mulhouse medal in 1859, a year when some of his fellow Englishmen still thought he was crazy. He recalled learning something from the visiting French professor St Claire Deville who tried a few experiments at the Royal College. 'One experiment was the casting of a large ingot of sodium, and the question was as to what vessel it should be melted in, when Deville noticed an iron tea-kettle standing by, and said that would be the very thing. This amused us very much; the idea of pouring melted sodium from a tea-kettle was indeed a novelty.'

The experiment was a success, but it was merely a dry run for a more prestigious display later that week at the Royal Institution. For this, the Institution's secretary, the Rev. Mr Barlow, thought a tea-kettle too undignified an instrument for such an august body. Accordingly, the sodium was melted in an open iron ladle, and the naphtha with which the sodium was covered burst into flames. Michael Faraday came to the rescue with water in a porcelain dish. 'If the tea-kettle had been used this would not have happened,' Perkin noted.

Perkin was delighted with all his trinkets, not least the still-wet portrait by Arthur Cope that he would hang in his home. It showed him in his Sudbury laboratory looking stout and proud in a velvet jacket, holding a skein of mauve wool; with his wild white beard he looked like a common vision of God. He promised to leave the portrait to the nation when he died. He also liked the bust by F. W. Pomeroy, to be placed in the library of the Chemical Society, the venue where Perkin reported the majority of his researches. When he joined in 1856, the Society's fellowship numbered 261; it was now more than 2,700. At the close of his speech, Perkin declared all this great acclaim to be particularly gratifying 'at this period of life, when the sun is declining in the west, and the evening is approaching . . .'.

Outside, in the Kensington streets, mauve was fashionable once again. In Brompton Oratory a wedding was taking place between Lord Gerard and Miss May Gosselin, and the six bridesmaids' gowns were of white chiffon, with mauve Empire girdles and mauve ribbon, falling at the back in long strands almost to the hem of their frocks. They each carried a white enamel Empire staff, topped by a ball of gold, and clusters of mauve Alexandra orchids and fragile asparagus fern were tied to the staffs with mauve satin. One newspaper called it 'the freshest cavalcade of colour'.

Inside the Hotel Metropole, a reporter from the *Daily Telegraph* was taking notes. He was working on the German angle, what he called 'the shadow-side to this epic of colour'. No one had yet written a more pained or stereotyped critique of Britain's failure to capitalise on its inventions, while making it clear that this was not a new affair. 'England had, as usual, in Sir William Perkin the genius. Germany had, as usual, a disciplined organisation in the shape of a host of trained chemists, capable of seizing an idea and working it out in all its applications.' The writer recalled Hofmann's prediction forty years earlier that nothing could prevent England from becoming the greatest colour-producing country in the world. But now England exported vast amounts of the crude by-products of its coal to Germany and Switzerland, and received back the finished article. 'We have forfeited our heritage, and upon the foundation of an Englishman's work the superstructure of the most commanding scientific industry in the Fatherland has been erected.'

Despite this, there was marvel. The *Telegraph* correspondent had recently been in India, where he had observed how Perkin had waved an invisible wand over 'the many-hued myriads' of the Asiatic cities.

> India is losing the more subdued and pensive harmonies, Old Testament-like in their effect, derived from the texture of its own looms dipped in its own vats. Everywhere the eye is notably struck by the more metallic and emphatic blaze of tints announcing the conquering march of the aniline dyes. Among the scenes of the great Delhi Durbar nothing is more vividly remembered to visitors to India than the astounding crash of orchestral colour . . . It was seen at once that the adoption of aniline dyes in recent years has raised Asiatic colour to a crescendo.

Some months after the jubilee celebrations, a local event took place in Perkin's honour in the Harrow region, with a musical note. A brass band played as Perkin and many Anglican friends strolled one evening in January 1907 to the New Hall in Sudbury, the mission Perkin had himself founded, to hear churchmen pay tribute to the inventor's devotion to spiritual and educational work. A certain Mr Wood, prominent in the Evangelisation Society, remarked how unusual it was, in that post-Darwinian age, for a man eminent in science to be also eminent in the love of the gospels of Christ. But there was much science in the Bible, he noted. Wood also praised the work of Lady Perkin, and it was his greatest privilege to present them both with an illuminated address from all their well-wishers at the New Hall and East Lane missions. The *Harrow Gazette* reported that, with the exception of gold, 'all the colours used in that illumination had been discovered by [Perkin], and he probably knew more about them than any other man in the whole world'.

Lady Perkin was given a silver tea pot as a souvenir. At the close she gave a little speech, saying that her husband had received many addresses and tributes, but this had been the only one in which she had been allowed to take part.

Perkin said he found all the celebrations exhausting, and was happy now to resume a normal life. The photographs of the period show no sign of frailty. When Ralph Meldola saw him in the middle of 1907 he thought that he looked healthy. Perkin had only recently returned from Oxford, where he had received the degree of Doctor of Science in the same ceremony as Mark Twain was made a Doctor of Literature. 'He bore all the excitement and fatigue without the least indication of discomfort,' Professor Meldola remarked. He had, however, picked up a virus of some sort, and there was a little creeping pain.

Perkin resumed his research into unsaturated acids until confined to bed on the morning of 11 July 1907. 'Although he complained of suffering pain he spoke hopefully of his condition,' Meldola noted, 'and anticipated being soon able to leave his room.' But it was a false note. Perkin had entered the arena of the unwell with double pneumonia and appendicitis, and these proved more serious than he or his family first realised. Near the end, Lady Perkin told her husband that they must be separated for a time. According to *The Christian* newsletter, his reply was: 'May you have much of the joy of the Lord.' An attendant then told him, 'Sir

William, you will soon hear the "Well done, good and faithful servant".' Perkin observed: 'The children are in Sunday School. Give them my love, and tell them always to trust Jesus.' He then let out the first verse of the hymn 'When I Survey the Wondrous Cross', and when he reached the last line, 'And pour contempt on all my pride', he said, 'Proud? Who could be proud?'

Then he fell asleep. Then he woke up, and at about six o'clock in the evening on 14 July 1907 he died, aged sixty-nine. His friends agreed that, in view of his activity, his death could not be regarded as premature.

The funeral at Roxeth two days later was a day of many blooms, most of them mauve. *Gas World* was there, and would tell its readers how the deceased gentleman's name was worked in mauve flowers on a white ground. The cortege consisted of twelve carriages, including one specifically for floral tributes, and Sudbury took off its hat and bowed its head as they passed. All the Perkins and all the chemists were there, with the exception of his eldest daughter Annie, who was in Tangiers when she heard of her father's death and got back too late. (His youngest daughter, Helen Mary, had been married in the same church just a year before, and Perkin had led her down the aisle to the tune of 'Father, I Know that all my Life is Portioned out for Me'.)

The grave was lined with lilies of the valley, and the coffin was solid oak and lined with lead. The brass inscription plate just gave his dates, but the tombstone that followed would be religious. The floral arrangements had notes attached: With loving sympathy, with deepest regret, to dear Grandfather with love from Wilfred and Isabel. One said: 'To Dear Lady Perkin, Please to accept these lilies from my poor little Charlie as he is so broken-hearted, and he wants to give these out of his garden as a last token of love with great sympathy. From Mrs Cattermole.'

At the service, Mr Wood drew comparisons between the life of Perkin and Enoch. The *Harrow Gazette* recalled four key points:

'(1)Why God took Enoch; (2) how; (3) where; and (4) for how long'. Not long afterwards, as the large crowds dispersed, the Sudbury Hall girls sang 'The Glory Song', Mrs Swaffield accompanying on the harmonium.

The following Sunday, J. W. P. Silvester, the vicar at the Wembley Parish Church, talked of Perkin's great inventions, and said he was proud to have known him. 'It is not the least pleasurable memory to me to know that almost the last letter, if not the last letter he wrote on earth, was, *inter alia*, an appreciation of my article on Funeral Reform, published in the Wembley Parish Magazine, July issue.'

His will left all his property to his wife. She also got his watches, jewels, clothes, plated articles, furniture, glass, china, live and dead stock, medals, musical instruments and £500. He left his servants ten shillings for every month he'd employed them. His children received a regular and equal annual payment, and his sons got his chemical apparatus and research books and specimens. The net value of his estate, which included over thirty freehold properties (and confirmed him as by far the biggest landowner in Sudbury), was valued at £73,444, but the sale of many buildings achieved far greater sums than those included in this estimate.

Among the many letters of condolence received by Lady Perkin was one from J. W. Bruhl in November 1907, who wrote from his university post in Heidelberg. 'In July, when reading in the paper the sad news of the loss you and indeed the whole world suffered at the demise of Sir William Perkin . . . The sudden death of your dear husband grieved me as much as it amazed me. Not long before he had written me a detailed letter in which he had expressed his satisfaction with his good health and spoke about diverse scientific matters. Only a few days before his end he sent me a copy of his last research with some especially kind words.'

And there was one from the ageing Heinrich Caro in January 1908.

Dear Lady Perkin,

It is a sincere comfort for me to know that the cruel bereavement which you and your dear family have suffered should not have broken the link which through a lifetime has connected my soul and mind with the immortal name of "Perkin", and that my thoughts may still continue to wander to the Chestnuts in Sudbury . . .

The attack of illness which so sadly interfered with my full appreciation of those wonderful days of the Perkin Jubilee was finally got rid of by a fortnight's sojourn in beautiful Folkestone. But of course the burden of my 74 years makes itself more and more felt, and the loss of nearly all my old friends and contemporaries causes an ever-increasing feeling of loneliness. My hardest blow was the premature passing away of Sir William, who had been my guiding star through my life and one of the best and greatest men of my time.

The obituaries for Perkin in the newspapers were full, and sincere. The *Daily Telegraph* wrote of his amiability and charming modesty. The *Daily Express* realised he had given a new industry to the world. The *Daily Mirror* was grateful that he had only been unwell for three or four days. The *Harrow Gazette* enjoyed the idea of Perkin's triumphs with coal-tar causing 'the deliverance from their chemical dungeon of the imprisoned spirits of the rainbow'. The *Daily Mail* estimated that there were now 500 colours that owed their existence to mauve. The *Manchester Guardian* wondered how anyone in those times possessed the genius to experiment on black sludge when most chemists would have thrown it down the waste pipe. It noted how well his intuition had rewarded him commercially. The *New York Times* recalled how Perkin had found fame in the 'black diamonds which the British Islanders dig from out of their coal mines'. In America, despite his recent visit, the Perkin story had also undergone transmutation: the chemist had been struggling with molecular formulae when 'one day, in a fit of disgust over the failure of an experiment, [he] made hot resolution to be quit of the science and seek other fields . . . [he] thought better of it the next day and lived to create more industries than most any who have ever gone before him . . . to open avenues for an army of workers in the arts of peace which numerically exceeds the standing army of Great Britain.' Back in Great Britain, *Tribune* wrote that Sir William's feat would remain unequalled until such time as chemists found a way of making artificial food.

For many readers this would have been the first they had read of this scientist. Perhaps some thought the eulogies were overplaying it a little, the way one always does when a good man dies. Few could have imagined how extensively he would be fêted in years to come, and how prolonged and significant would be the effects of his work. Or how, in subsequent years, Perkin would also be blamed for the perilous state of the industry he had himself created.

Five miles from the heart of Bristol, on the site where food scientists first made the drinks that were marketed as Ribena and Babycham, two biological scientists are developing ways of reintroducing ancient woad. In a glasshouse at the edge of the Long Ashton Research Station, David Cooke and Kerry Gilbert are growing 300 small plants, none more than a half-metre tall, and are employing modern methods to extract indigo for textiles and inks. Theirs is an expanding market – offsite, they have farmers working 15 hectares to meet their demands.

In the second week of January 2000, Dr Cooke was in his laboratory telling a Belgian visitor of the historical significance of his endeavours. 'In Queen Boudicca's day, the army used to smear their faces with woad to look more fearsome.' Although he has no great evidence, Cooke likes to believe that woad was also an antiseptic, and that smearing before battle could guard against infection from wounds.

Cooke is a wiry man of fifty-three, with a boyish demeanour. He has worked at Long Ashton for thirty years, specialising in the regulation of cell-membrane stress in plants. He has no great liking for many of the products or methods of modern chemical industry.

He says that the philosophy behind his work on woad is three-fold. A lot of the materials we use today, including dyes, are derived from the petroleum industry, the successor of coal-tar as the major source of chemicals. But of course these are not infinite resources. As the supply declines, so the cost will go up, and eventually governments will prioritise their usage. This will result in a decline in our standard of living. As Cooke says: 'If you want to go on a holiday to Spain but you can't get on an aircraft because there's no fuel, you may get a little bit upset. If you want to go to London in a car and you can't, you're going to get a little bit upset.' There are thousands of aspects of one's life which may be affected by the decline in fossil fuels, although Cooke is reluctant to estimate when this day will come.

To prevent too many people becoming too upset, he has, with the aid of a European Community grant, been looking for alternative sources of chemicals. Dyestuffs were chosen because they were the most obvious and easiest: if you want to derive a colour from a plant it's a great deal simpler than trying to extract a great many other esoteric molecules. And of course it's been done in the past.

The woad project would also benefit farmers. With profits down, and a European surplus in many meats and crops, farmers are leaving the land. 'This is a bad thing,' Cooke maintains, 'so they have to be provided with an alternative to food production, and industrial crops seem to be the answer.'

The third element of Cooke's philosophy attempts to provide the customer with choice. A lot of people prefer to wear natural textiles rather than polyester or nylon, but they find that the artificial dyes on natural textiles are irritants. 'It spoils the whole thing. It used to be that the washing powder industry had a lot of problems with the allergies caused by their powders; they have now improved their product to the extent that there are very few allergies, but new allergies are still being traced to synthetic dyes. There is an ever-growing list of dyestuffs which are banned for use on textiles. So by providing natural dyes we can supply a niche market.'

Above Dr Cooke's head is a board with illustrated examples of dye plants grown in the UK: woad, dyer's weld, madder, yellow camomile, golden rod, greater celandine (yellow). Woad was chosen because there are 80,000 tons of indigo sold in the world each year, representing about 10 per cent of the world's dye market, most of it going into jeans. So there is a definable market, and it would be easier to persuade the dyers to take a natural product if they knew there was a demand for it.

'It's not as if we're producing yellow,' says Kerry Gilbert, Cooke's research associate, aged thirty, probably the only person in the country to have a PhD in the Cultivation of Woad. Their first indigo project lasted from 1994 to 1997, at the end of which they were able to show that it was both financially and industrially feasible to produce the dye naturally again. Then they were commissioned by the Ministry of Agriculture, Fisheries and Food to make a natural indigo ink for computer printers. At present the computer ink market is dominated by inks made from methyl ethyl ketone, a chemical that's allergenic and carcinogenic. The Ministry was concerned that the ink used to print the sell-by date on the labels of perishable goods may be absorbed into the food, particularly in those products with a high fat content that take up solvents at a very high rate. Also, Cooke says, 'computer inks are the only product on the market at the moment that are worth more than their weight in gold. That's how the computer companies make all their money – the printers they sell you are dirt cheap, but they fleece you on the ink cartridges. Although it's a niche market, it's an extraordinary profitable one.'

Cooke shows his visitors the results of his work – a large flask with a little blue ink sloshing inside, and another with an oily yellow solution that will only turn blue when in contact with the air.

'It's worked very well from all points of view,' says Dr Gilbert.

'We've developed a better method of extraction, and in the lab we've done some rudimentary breeding, and some good molecular work looking at the genetics that control the colouring features.' She has calculated that 200,000 hectares of land are required to produce enough woad to supply the UK's demand for blue.

'People do tend to think of natural dyes as only being these washed-out browns and beiges and greens,' says Dr Gilbert, who is wearing a sweatshirt embroidered with the slogan United Colors of Benetton.

She produces a box containing wools and cottons dyed with natural indigo, some in a deep and sumptuous hue. She has hooked up her Windows screensaver to run a pattern of multiplying pipes coloured with a photograph of these natural colours. 'It takes a bit of persuading to get people to accept that you can get almost the same array of colour from natural sources, not least because the chemical industry has helped us understand the formula of dyes far better. We can now use what chemistry has learnt, but in a natural way. This knowledge wasn't available in the nineteenth century.'

She explains the process. She takes the leaves and extracts them into water. She uses hot water to burst the cells, and the precursors (chemical substances that give rise to more significant ones) are released into the extraction liquid. She then cools it down, and alters the pH to make it alkaline (by adding lime). Then she aerates it and pursues the ancient process of drying and fermentation. In the traditional method there was much wastage, as the precursors would frequently react with other molecules and form other by-products; with the modern process, the conditions are altered chemically to ensure that the base molecules making indigo generally do meet another suitable molecule. As indigo is insoluble it sinks to the bottom of a tank, making it easy to dispose of the top layer of liquid and leave basic indigo, which is then dried into powder.

'Obviously we're not going to replace all synthetic chemicals,'

Gilbert says, 'but I think people are pleased that the balance is moving back.'

'This puts the foot in the door,' Cooke continues. 'It suggests that it is possible for life to continue as we know it after coal and oil is gone, that we have not completely forgotten how to produce things in a way that is not artificial.

'It's like all these things. It will start as a trend. Some designer will feature naturally dyed garments as part of their Paris collection, and then this will be taken up by the well-heeled, and then the fashion chain stores will make copies, again with a natural tag on them, and then everyone will latch onto them.'

And there is another route for the re-emergence of woad. 'Levi's have already started making custom jeans – you ring up or go to their Internet site and give them your measurements and choose your style of cloth and type of fade that you like. It may cost anything from £100 to £200. And one of the options which Levi's would be very keen to offer will be a naturally dyed pair of jeans. For the wearer it could become a source of pride, perhaps a sort of one-upmanship. It may be considered a slightly greener alternative, and for that they will be prepared to pay a premium. For Levi's, jeans dyed with natural indigo may only cost 50 pence more per pair, but they could charge £20 more. Once that starts to take hold as a fad, it will gradually find its way into the mainstream. Recently there was a fashion for army-style combat trousers, but that's passed. People are going back to blue jeans again. And Levi's have told us that as soon as we can provide enough natural indigo to satisfy their requirements, they'll buy it.'

11
SELF-DESTRUCTION

Like William Perkin I personally aspire
to metamorphose lower into higher.
His transforming coal-tar into brilliant dyes
has come for me of late to symbolise
chemistry at its most profound and true
creating radiance out of basest residue.
It is infinitely satisfying
to see what was black dreck serve the art of dyeing.

Those are the powers and forces that are needed
if the Western World is not to be superseded.
William Perkin's ingenious transformation
could have benefited the British nation
But unfortunately Great Britain still relies
on Germany for all synthetic dyes.
The coal tars of the Ruhr and Rhine
metamorphosed in industrial bulk into aniline.
If more of us in Britain had done work in
the processes pioneered by William Perkin
we would not now as a nation still have to rely
on Germany for all synthetic dye.

Like many a physical or chemical invention
pioneered by the British I could mention
Perkin's valuable synthetic dyes

which will always, for yours truly, symbolise
the magic of chemistry, Germans monopolise.

Sir William Crookes, historian, physicist, spiritualist and chemist (he trained under Hofmann, discovered the element thallium and conducted important experiments on cathode rays and radiation), as portrayed in the play *Square Rounds* by Tony Harrison, first performed at the National Theatre in 1992

In December 1914, William Perkin's third son, Frederick Mollwo Perkin, delivered a speech to the Society of Dyers and Colourists which suggested that a particular problem in the dyestuffs trade was having a dramatic effect on the war effort. But the meeting at which he spoke was sparsely attended, for many managers of English dye companies were engaged in an emergency meeting at the Board of Trade to ensure an adequate colour supply for the manufacture of soldiers' uniforms.

In Germany, the country which now supplied 80 per cent of Britain's artificial colour, the massive dye factories were being expanded to make a product which used the similar basic materials of benzol, toluol, nitric and sulphuric acids: explosives.

'A devastating war has broken out, stopping our supplies of imported colours, and what is the result?' Perkin asked. 'There is a dye famine. Dyers cannot carry out their contracts because, although willing to pay almost any price, they cannot obtain the dyes.' The value of imported dyes was estimated at between £2 million and £3 million per annum, and the £100 million textile industry was entirely dependent on them. Other industries were also badly affected: paper, leather, bones, paint, wood and food increasingly relied on synthetic colour. At the start of the war, British manufacturers produced only about 15 per cent of all dyes used in the United Kingdom, and although the priority had switched from fashion to the military, it was clear that a once-great industry was now failing its people. The fact that wartime apparel was drab is usually explained in terms of frugality and respect; in fact, there were simply no colour dyes. And the relatively small scale of the dye industry in Britain meant that the supply of nitro compounds needed for explosives was quite insufficient. What had caused this crisis? Perkin's explanation was extensive and damning. At its heart lay an attempt to defend the reputation of his father.*

'I have seen it in the Press,' Dr Mollwo Perkin announced, 'and have also heard it in conversation, that it was not a very patriotic step to dispose of a successful works at the age of thirty-six.' But he reasoned that his father had no choice. As with other industries, the dye trade required continual change: only the companies with the best new colours, and the most efficient methods of making them, would capture the market. 'It was not possible for one brain, however energetic and fertile, to carry out all the necessary research required . . .' And research chemists could not be obtained because British universities were reluctant to train them. There were German chemists of course, but most who gained experience at British firms soon returned home. Britain did not foster a good relationship between its universities and industry, and industry had nothing to compare with the magnificent research laboratories in Germany and Switzerland. At the outbreak of war, Hoechst employed more than 200 research chemists, where Levinstein and Read Holliday had but a handful. And in Germany they had been building on a green field, with fewer and smaller established industries telling them which way to go. New industries developed on the Rhine without having the

* After the war, Mollwo Perkin became custodian of some of his father's property and what remained of his dyestuffs. On 24 October 1922 he wrote to H. E. Armstrong, a former student and historian of the Royal College of Science, from his home in New Oxford Street. His telegram address was Mauvein, Westcent, London. 'I have pleasure in sending you a sample of the Original Mauve made by my Father in 1856 or 1857. The bottle which I have is marked 1856 so I presume it was made in that year. It certainly was not manufactured at the Factory.

'I have very little more left – but I am sure you will appreciate it. I also have pleasure in enclosing a small piece of silk which is a portion of a dress accepted by Queen Victoria at the 1862 Exhibition.'

weight of a hundred years of textile industry on their shoulders.*

But there were other reasons for Germany's dominance, and they pained Perkin to reveal them. 'There certainly has been a lot of piracy,' he claimed, citing the chaotic British patent laws. In fact, in the first decades of the industry these laws were so haphazard as to offer hardly any protection at all, and hence chemists from Berlin and Munich appropriated the best brains and foreign methods for their own use. The British government allowed foreign chemists to take out patents in Britain which they were not required to work, but German patents were invariably refused to British inventors.

Perkin had other explanations, which soon began to sound like excuses: British industry received far less outside investment from banks and other institutions; British capitalists lacked the long-term vision to reinvest in new processes and machinery, and chose immediate profits over continued research; British capital received a greater profit from shipping, the docks and the mines, or abroad with the advance of the American railroad. Then there was the problem of marketing skills: armed with huge colour sample books, German dye salesmen had the confidence to convince anyone that they could supply any shade cheaper. And then there was also the question of alcohol: the large quantities of pure alcohol required for

* C. M. Whittaker, a senior dye chemist at Read Holliday in Huddersfield, recalled how, during the war, those in the dye trade went from being relatively unrewarded and unprotected into the most sought-after of professionals. 'The period of the First World War was a severe moral testing time . . . One became a little tin-god on wheels overnight: one was fawned on from all sides.'

Read Holliday (as part of British Dyes Ltd) supplied khaki yellow and khaki brown. 'I have frequently said that the then-small British dyemakers have never been given proper credit for the competent way in which they fulfilled the demand for dyes for the uniforms of the British and some of their Allies. The shortage of dyes was such that the prices obtainable made it frequently more profitable for a dyer to sell his actual dye stock rather than use it.' At the end of the war the British Dyes staff in Huddersfield were fêted with a lavish dinner-dance, 'during which a bombastic speech told them that they had helped to win the war. On the Saturday morning over fifty chemists got the sack.'

some of the newer dyes was made prohibitively expensive in Britain by duty and excise restrictions, whereas in Germany it was plentiful and cheap.

Perkin's analysis carried an air of desperation. The war had cruelly exposed British weakness, but the danger had been gravely apparent when his father was still alive. Indeed at the jubilee celebrations of mauve, Richard Haldane, the Secretary of State for War, acknowledged the criticism pitched towards his cabinet. Haldane was the key speaker at the Hotel Metropole, and he began his address looking towards the guests from abroad. 'I have often thought that it must strike you as somewhat odd to see the way in which we attend to these things in this country.' One of the British chemists then shouted out, 'Or do not attend'.

Haldane said he was thrilled with the globalisation of the industry. He noted the 'magnificent organisations' of the dye industry abroad, 'and the extraordinary capacity your governments show in excess of ours in taking care of science'. He had a reason for the different approach in the United Kingdom. 'We are a very practical nation,' he said, 'and yet we do muddle through somehow.' People in his audience laughed. 'It is quite true that our state takes, in the first instance, but little interest in science.' This was an extraordinary admission from a minister. His reasoning suggested that there was no need for any great institutional support, for genius appeared to do it all by itself. 'We do not organise in this country, we hate abstract ideas, and we put difficulties rather than encouragement in the path of the discoverer.' Haldane said that the great scientists and architects and engineers still made their way to the front, and that night he had one prime example: the mention of Perkin's name inevitably drew applause, but Perkin himself must have looked away.

Haldane continued with an interesting analysis, in which he blamed the calamitous state of the colour industry on the death of Prince Albert. 'I have often thought that if the Prince Consort had lived, probably Hofmann would have remained in this country, and . . . aided by that great spirit, the centre of the coal-tar industry and all its products and – what appeals to us so much – of the many millions which have come out of those products [laughter], would have remained in Britain and not have passed to Germany.'

What a shame, the minister noted, that British universities did not resemble factories with great chimneys, the way they did in Berlin, or house twelve professors of chemistry. He spoke as a War Secretary in peacetime, and could have little known the far-reaching consequence of his words just seven years on.

A short while later, it was left to Carl Duisberg, the general director and an expert marketeer at Bayer, to explain why his country now put the British colour trade to shame. It was not lack of finance: Britain was still the wealthiest country in the world, while Germany, when it began its organic chemical industries in earnest just thirty years before, was one of the poorest. He claimed that patent laws, or the lack of them, had not been the main key to German expansion.

Duisberg claimed there were more fundamental ethnic characteristics at play, one of those potent generalisations that surface most frequently at a time of national tensions. The English were simply not good at patience and the sort of self-belief required when waiting for success. 'For all that the Englishman does he expects soon to be compensated in hard cash,' Duisberg claimed. He was perhaps thinking of conservative British bankers' willingness to let their foreign counterparts invest in high-risk scientific endeavour; there was simply no way of calculating the potential earning capacity of chemical discovery. 'It requires above all a singular ability to wait and bide things coming combined with endless patience and trouble,' Duisberg noted. 'We Germans possess in a special degree this quality of working and waiting at the same time, and of taking pleasure in scientific results without technical success.'

The prime example was indigo. The biggest single factor in Germany's success lay not in Hofmann's defection, but in Baeyer's breakthrough, and that had come about after two decades of furtive and intensive research involving hundreds of scientists from several large companies. There was little doubt that ultimate success would bring riches and acclaim and a large springboard for further technical advance, and so it proved for BASF and Hoechst. In 1905, two years before the death of William Perkin, Baeyer's work on indigo won him the Nobel Prize.

Carl Duisberg argued that as soon as Germany had the monopoly on indigo there was not much that could be done. Britain had suffered in India, but this was nothing compared to how it would suffer at home. Within ten years of the first commercial manufacture of indigo along the Ruhr, Germany's output was probably around 100 times the total 60-year production of mauve, and fast approaching the total output of all the other artificial colours produced to date.

English dye companies put up disciplined resistance. Their strongest voice was German, and came from Ivan and Herbert Levinstein's once-flourishing factory at Blackley. Levinstein's own ambitions to produce indigo were thwarted by that peculiar facet of English patent law that allowed foreign firms to take out a patent with no obligation to work it. In this way overseas companies sealed their monopolies: Levinstein claimed that out of the 600-odd British patents assigned to foreign companies between 1891 and 1895, not one was being exploited.

What really rankled was the destructive sense of German arrogance (which may now be seen as prescience). In 1900, Heinrich Brunok, managing director of BASF, suggested that the production of Indian indigo should be abandoned in favour of food crops. The British government shrugged by with half-measures, including the ruling that military uniforms should only be dyed with natural dyestuffs, and by the time it woke up to the scale of the problem in

1907 Britain's competitors had established a wholly intimidating advantage.

In this year, David Lloyd George, President of the Board of Trade and the man later honoured by ICI as the person who did most to rescue the British chemical industry, reformed the patent laws. From now on, overseas companies were required to work their British patents or risk losing them, an amendment that speeded up the establishment of the Hoechst indigo works at Ellesmere Port, near Liverpool, the following year. One of its barrels of indigo was sent to Lloyd George with a message embossed in gold: 'Made in England'.

But profited by Germans. Hoechst (which was still strictly known by the name of its founders, Meister Lucius & Bruning) employed predominantly German workers who operated under intense security. Levinstein complained that even when German firms revealed their patents, they deliberately omitted vital information, and concealed the fact that one key ingredient, phenylglycine, was wholly imported from factories by the Rhine. Besides, although the production of indigo at the Ellesmere Port works increased from 9 tons in 1908 to 293 tons in 1913, this was quite insufficient to meet British needs. In 1913, 1,194 tons were still being imported from Germany, and this despite the recent establishment of a BASF factory on Merseyside. In 1914, the three leading British firms Ivan Levinstein, Read Holliday and British Alizarine together produced about 4,000 tons of dyes. The major German firms produced 140,000.

Before the outbreak of war, the Germans proffered one damning response to the ambitions of the British dye trade: relax and retire. Other nations should not envy Germany's position, but simply leave it alone. Carl Duisberg believed that England really had 'no cause for complaint about her success and position in the world, and especially no cause for complaint that perhaps one or another country has superseded her in one or other industry'. England had highly developed coal and iron industries, and the leading spinning and weaving factories. Before the war, England had more colonial possessions than anyone. But only in the coal-tar colour industry must she be satisfied with second (or third or fourth or fifth) place behind Germany (and then Switzerland, the United States and Russia). 'Why should Germany in this one instance not take a leading position?' the man at Bayer wondered.

It was argued that the German dye industry was already too huge, and enjoyed such spectacular dominance, that its position was unassailable; even decades of inspired competition would only chip away at the edifice. As it turned out, war would make a little difference, but it was an intriguing proposition: so what if England invented cricket; even a proud nation would come to accept that other countries would one day learn to beat them at it every time.

All of which might have made good sense, and would not have mattered beyond repair (just another – not the last – example of British genius shipped abroad by money and ignorance), had it not been for the fact that the dye trade was not just the dye trade any more. By 1914, at the beginning of the first industrialised war, the chemistry that made dyes had advanced to the stage where it could now alleviate pain and save lives. William Perkin's initial fumblings with the formula for quinine was bearing fruit in unfamiliar ways (though still not with artificial quinine, which would not be synthesised until the next war). Perkin had shown in his lifetime that once you could make mauve and alizarin, you could then also make artificial perfume and saccharin. These were complex conquests, but they represented a logical advance to most skilled research chemists. The more these chemists were valued and rewarded, the greater would be the eventual benefits. What was needed was the prepared mind. And it was inevitable that the next major chemical advances would occur in Germany.

Because of the huge share they enjoyed of the world market, the six major German companies were soon at each other's throats.

With such riches to be made, competition was fierce. This led to the undercutting of prices and the further decimation of rival markets, but also to a reduction in profits. The solution lay in the formation of two loose cartels, each of which carved up a larger slice of the market and ensured efficiency of production and agreements on prices. In 1904, Hoechst combined with Leopold Cassella (Frankfurt) and three years later with Kalle of Biebrich. Bayer joined forces with BASF and AGFA, a group that became known as the Little IG (an *Interessengemeinschaft*, a community of like-interests). The man behind these collusions was Carl Duisberg, who had visited the United States in 1903 to set up a Bayer factory in New York and there learnt about the booming trust movement, and in particular John D. Rockefeller's Standard Oil trust.

The template worked well in Germany, and put an end to price cutting and patent swindles while retaining each company's autonomy. It also meant that the firms were freed up to develop their other chemical interests and thus ensure their expansion.

AGFA became the largest European manufacturer of photographic materials. Bayer made aspirin, the worlds biggest-selling pain reliever, manufactured since 1899 from the dyestuff intermediate salicylic acid. It also made heroin and methadone, and, with a poignancy that would have delighted William Perkin, its dye money also funded the successful launch of Atebrin, a new malarial treatment. BASF led the search for a synthetic ammonia that would free Germany from its dependence on the monopolistic supply of the natural fertiliser from Chile. It had another use – as the base of almost all modern explosives. Both applications depended on the synthetic formation of ammonia, a breakthrough attained after a massive effort by Fritz Haber, who successfully combined nitrogen and hydrogen (a process developed commercially by Carl Bosch at BASF in 1913 and still in widescale use). In 1915 Haber was the first to develop chlorine as a large-scale war gas.

At Hoechst am Main, much dye money went to support the

work of Paul Ehrlich, the brilliant scientist who used William Perkin's colours to change the course of medicine. Ehrlich studied in Breslau and Strasbourg, and took his first job at a hospital in Berlin. Here he worked on typhoid fever and tuberculosis, a disease he contracted himself. He was a small, wiry man, who took to smoking 20 cigars a day. From his earliest studies he was convinced of the limitless potential for using coal-tar and other dyes in the development of biology and medicine, and his conviction led him to pioneer the earliest forms of chemotherapy.

The staining of cells with aniline dyes was first conducted in the early 1860s by F. W. B. Beneke of Marburg (who used mauve) and Joseph Janvier Woodward, a surgeon in the US Army, who used fuchsine and aniline blue in his examinations of human intestines. Two factors made this possible. The first was a great technical advance in the microscope, which, invented in 1590 by a Dutch spectacle maker, consisted initially of a magnifying lens set in a wooden frame. The English physicist and architect Robert Hooke, who helped plan the reconstruction of London after the great fire of 1666, used an improved microscope to first describe the presence of compartmentalised elements he had detected in cork, thus giving rise to the concept of the living cell. Great advances were made by the Delft draper Anton van Leeuwenhoek who left 247 microscopes when he died in 1723, many with fixed samples of the 'little animalcules' (protozoa) he had observed in rainwater and semen. He employed one of the earliest uses of natural dye staining when he improved the visibility of the muscle fibres of cows by marking them with a yellow saffron solution. The Englishman Sir John Hill subsequently used extract of logwood to study the microscopic structure of timber. And a few years before the discovery of aniline dyes, Joseph von Gerlach used carmine to make an important analysis of brain specimens. Many of his microscopes used a primitive screw to focus, and results were limited by the impure quality of the lenses and strong chromatic distortions. Improvements came gradually – the variable eyepiece, the elimination of the aberrant deflection of coloured rays, improved glass lenses – until by 1869 it had evolved sufficiently to enable crucial early advances in genetic research.

In this year a Swiss chemist called Friedrich Miescher used aniline dyes in his detection of 'nuclein', that part of a cell nucleus that was not protein. Nuclein contains phosphorus, and was later renamed nucleic acid. One form of this, deoxyribonucleic acid, we now know as DNA. After his great discovery, but perhaps without realising its implications, Miescher then spent the rest of his life studying patterns of fertilisation and the nuclein in the sperm of German salmon. It was left to such men as Walther Flemming to use the most basic aniline dyes to perceive the threadlike structures in the cell nucleus that would later be called chromosomes. Flemming developed early theories about cell division with his work on salamanders, and coined the word 'mitosis' to describe the splitting of chromosomes along their lengths into two matching halves (the word 'chromatin', describing the vivid colour within a nucleus after staining – derived from the Greek word for colour, chroma – was also his).

It was Paul Ehrlich who realised that the chemical dyes obtained from coal-tar did not simply colour cells or tissue samples with their own tint, but often combined with a substance to form a definable chemical reaction. The aniline colour methyl green, for example, while leaving the nucleus green, would stain the cytoplasm of a cell red. In 1875, Ehrlich's cousin Carl Weigert had demonstrated that the fuchsine derivative methyl violet stained bacteria in tissue samples (as opposed to the tissue itself), and it was probably this observation that inspired Ehrlich to devote much of his early career to the new science of staining, and confirm its key role in the identification of nucleic acids, sugars and amino acids.

His first significant impact was felt in the Berlin laboratories of the bacteriologist Robert Koch, the Prussian who had found fame with his discovery of the bovine anthrax bacillus and his theories of how germs might spread between animals and cause disease in humans. Koch had studied Ehrlich's latest staining techniques, and used the aniline dye methylene blue to detect and then prove the existence and effect of the tiny rod-shaped bacillus in the tissue of those struck down with tuberculosis. Similar work with cholera followed, and in this way did Koch make huge strides in our modern treatment of disease.

Paul Ehrlich claimed that the evening in 1882 on which Koch announced the cause of tuberculosis was his greatest experience in science, and it inspired him to work with Koch on the development of the drug tuberculin (an only moderately successful treatment, but an effective indicator of disease). This work led to Ehrlich's own practical discoveries at the dawn of what would soon be referred to as biochemistry, the fusion between chemistry and physiology. Ehrlich's work also relied on another collusion – that between medicine and chemical industry. Even though the molecular structures of many colour dyes were being unlocked with great frequency in the 1880s, Ehrlich would not wait for exact formulae to begin his own experiments. New textile colours were immediately incorporated into tissue and cell staining, and some of the results were extraordinary.

Methylene blue was found not only to be effective as a diagnostic in bacteriological work (including malarial cases that were not responsive to quinine), but also to possess properties that were themselves intrinsically useful in medicine. It was found to be a mild antiseptic, and it was one of several coal-tar derivatives to be important in Joseph Lister's development of antisepsis and sterilisation. The ability of the dye to transform haemoglobin into methaemoglobin (in which the iron has been oxidised and plays no part in oxygen transport) was used to treat cyanide poisoning, since methaemoglobin turns cyanide far less toxic. Methylene blue was employed in Ehrlich's pioneering studies of living cells (as opposed to previous work on animal and human cadavers); his injections of the dye into frogs vividly stained the nerve cells, an observation of invaluable use to anatomists' studies of the nervous system.

In time, the chemical properties of several other dyes were also found to have significant therapeutic results and became listed as official drugs. Congo red was found to be therapeutic as a treatment for infectious rheumatism and an antitoxin against diphtheria. Scarlet red has been found to stimulate the growth of specific cells and for the care of chronic ulcers and burns. Acridine yellow was used as an anti-bacterial agent from 1916, while the orange-red fluorescein dye mercurochrome was widely used as a disinfectant for small wounds. Another orange-red dye known as Prontosil red was found by the German biochemist Gerhard Domagk to be anti-bacterial (it combated streptococcal infections), and led to the development of important sulfa (or sulphonamide) drugs which fought off puerperal fever, pneumonia and leprosy. Gentian violet was employed for antibacterial and antifungal purposes.*

It was the widespread use of dyes in medicine and pharmacy (in the colouring of pills and mixtures) that necessitated a new standardisation of the descriptive names for colours. By the time such an index was established in the United States in 1939, eighty-three years after mauve, it contained the names of just over 7,500 synthetic colours.

By a nice twist of fate, Ehrlich gave something back to textiles in

* In 1917, largely as a result of experiments at Levinstein and Co. at Blackley, the *Manchester Guardian* reported that from this time on, 'dyes and drugs must be thought of together. Whatever serves the modern dyemaker directly serves national health.' Domagk's work on Prontosil and the sulfa drugs in the mid-1930s is credited with prolonging the lives of Winston Churchill and Franklin D. Roosevelt Jr, and was considered so important by Alexander Fleming that in the early 1930s he temporarily prioritised work on Prontosil over his development of penicillin. Prontosil's formula was tweaked by the French and the chemists of May and Baker in Britain, leading to new so-called M & B drugs before the war. Domagk's work led to the Nobel Prize for Physiology or Medicine in 1939, but Hitler's disapproval of the award of the 1935 Nobel Peace Prize to Carl von Ossietzky, whom he had imprisoned, led to Domagk being unable to accept his award until 1947.

the form of new dyes. His work on methylene blue showed that dye was transported through the bloodstream, and entered the cells as fine particles. He determined to discover whether this process was due to the particular colour, or to the sulphur it contained, and so substituted the dye with one in which the sulphur was replaced by oxygen. In this process he collaborated with Heinrich Caro, and Caro's own search for a substitution yielded the new class of rhodamine dyes, which are still widely used in biological staining.

Ehrlich's lasting fame came from his work on blood cells and immunity, for which he won a Nobel prize, and for introducing Salvarsan (1910), the synthetic chemical that treated syphilis (which was also called 'Ehrlich's 606th', a reference to the fact that Salvarsan had been discovered only after he had tested but discarded another 605 similar compounds). Initially, Ehrlich's achievement with Salvarsan received little acclaim, as those with syphilis were widely thought not to merit a cure. But with time it was acknowledged that this dye work established the viable search for what he termed 'the magic bullet', a process that involved first highlighting and then targeting specific disease-causing micro-organisms by altering the chemical structure of staining molecules – the foundation of chemotherapy. Before the introduction of penicillin in the 1930s, and many years before the principle was used to treat cancer, chemotherapy drugs served as the main (albeit limited) treatment of septicaemia, pneumonia and meningitis. Hoechst also developed Novocain, a revolutionary synthetic local anaesthetic still employed in dentists' surgeries. The success of Salvarsan also led to new therapies for tropical diseases, including mepacrine and proguanil, two of several new drugs which, between the wars, began to supersede quinine as the most common treatment of malaria.

The Great War transformed colour throughout the world. It was inevitable that trade bans and isolationism would force all countries that had relied on Germany for its dyes to find other supplies; it was remarkable how swiftly and successfully this challenge was met. In Britain, the war served to resuscitate the entire industry.

The solution lay in takeovers and mergers. At the government's behest, Ivan Levinstein and his son Herbert took over the Hoechst indigo works at Ellesmere Port, while the BASF factory at Birkenhead was taken over by Brotherton of Leeds.

Read Holliday merged with the Bradford Dyers Association and Calico Printers Association to form British Dyes Ltd, a company that then merged with Levinstein Ltd and smaller companies in 1919 to form the British Dyestuffs Corporation. Jointly, these companies supplied not only enough dye for uniforms and other products, but enough intermediate substances such as the nitro compounds required for the propulsion of shells. The success of this enterprise – based on the hard-won realisation that the development of organic chemistry actually had a significant role to play in any modern society – would inform the establishment of Imperial Chemical Industries in 1926, a merger between United Alkali Co. Ltd, Brunner, Mond & Co., Nobel Industries Ltd and the British Dyestuffs Corporation that swiftly accounted for 40 per cent of all chemical production in Britain. ICI also incorporated the British Alizarine Company, the London firm that had swallowed up what remained of the supplies and patents of Perkin and Sons, and in this way could trace its lineage back precisely seventy years to the discovery of mauve. By the Second World War, the lessons learnt at the start of the Great War ensured that Britain was self-sufficient in shellfire and the ability to dye its own uniforms.

In the United States, where the reliance on German dyes had been as crippling as in Britain, a similar transformation occurred. Addressing the National Silk Convention in New Jersey in 1916, the chairman of W. Beckers' Aniline and Chemical Works noted that any country that had its national defence at heart should possess a strong dye-making industry within its own borders. Beckers noted a phenomenal growth even before the United States entered the war: in 1914 only five factories were actively engaged in the production of aniline dyes, while just two years later over eighty firms dealt in coal-tar products – either with intermediates of colour dyes or dyes themselves. Partly this was down to national pride, and what was seen as a willingness on the part of the consumer to accept slightly unpredictable colours if they were made in America. Dr Beckers told his audience of the particular problems he was having with putting methylene blue on silks: 'If you would not have been broadminded, and would not have taken from our hands ton lots after ton lots of such dyestuffs which were not quite up to standard shade, we would have gone bankrupt at the start.' It helped that the war ensured that consumers had no choice in the matter.

In 1916, Britain played a significant role in the establishment of the Du Pont dyeworks, with Herbert Levinstein exchanging information on the production of indigo. In 1914, the American dye industry employed just 214 chemists; by 1919 the number was 2,600.

In Switzerland, which had greatly increased dye production during the war and supplied the majority of British imports, Ciba, Geigy and Sandoz jointly formed the Baseler Chemische Industrie, while similar cartels appeared in France and Italy. In Germany in 1916, eight companies formed a conglomerate to prepare for greater competition in a post-war market. The Interessengemeinschaft der Deutschen Teerfarbenfabriken – IG Farben – included BASF, Bayer, Agfa and Hoechst, a syndicate that instantly became the biggest chemical company in the world, and would soon try to take it over.*

* IG Farben's official formation came in 1925, by which point it had incorporated 50 dependent firms and controlled more than 100 plants and mines. At its peak in the inter-war years, IG Farben employed 120,000 workers, including 1,000 qualified chemists. Its dyetrade accounted for about a third of its output, and although it did not produce the best new dye innovations in this period (that accolade went to ICI), its huge expansion was due to its skilled marketing experience, especially as it switched from its closed European markets to the East.

12
THE NEW EVENTUALITY

Colour may be extracted from substances, whether they possess it naturally or by communication, in various ways. We have thus the power to remove it intentionally for a useful purpose, but, on the other hand, if often flies contrary to our wish.

Johann Wolfgang von Goethe, *Theory of Colours*, 1810

He, Mertens, was a young chemist, German and Catholic, and I a young chemist, Italian and Jewish. Potentially two colleagues: in fact we worked in the same factory, but I was inside the barbed wire, and he outside. There were forty thousand of us employed in the Buna Works at Auschwitz. That the two of us, he an Oberingenieur and I a slave-chemist, ever met is improbable . . .

Primo Levi , *Moments of Reprieve*, 1981

Lady Perkin died in the spring of 1929 at the age of ninety. After her husband's death she had maintained his charitable work for local church groups and the Salvation Army. Her daughters had made her proud, one taking up missionary work, another becoming a nurse. At the time of her death the local paper called her 'a kindly, gracious and lovable Lady Bountiful', and observed that Sudbury had quite transformed itself from the once secluded village she knew at the time of her marriage. By 1929, there were very few locals who could tell you much about her husband's pioneering endeavours, or what had become of them.

Nine years later, 100 years after his birth, Perkin was a celebrity once more, at least in the halls of the chemical societies. On this particular anniversary, dye people made speeches, the past was exaggerated, bow-ties were worn. At the London patent office, there were the clear signs of a resurgence in the British colour industry, although the frequent presence of the letters IG emphasised the continued and towering force of German competition. The patent specifications included new dyeing applications – coloured moulded articles, hardened casein, cellulose acetate rayon – new colours used in new ways on the new plastics and man-made fibres. And on the East Coast of America, Du Pont had just registered a new material called nylon.

As with most anniversaries, the centenary of Perkin's birth may teach us useful lessons in reputations. We may learn, for example, that William Perkin was now revered for lighting the path towards rayon, synthetic rubber and other items indelibly linked with a modern age. When dye chemists got up at the Leathersellers' Hall and the Victoria Hotel, Bradford, to sing of Perkin, their audience found plastic on their tables and another story of amateurish scientific romance.

Belgian Leo Baekeland shared the occasional career twist with William Perkin, and some with the Mad Professors of folklore. Baekeland trained as a chemist in Ghent with an ambition towards academe, but a honeymoon in America made him aware of how industrial chemistry might meet the demands of a consumer society. In 1899 he made $1 million dollars by selling George Eastman his invention of Velox, an improved photographic paper that enabled users to develop their work in artificial light.

Baekeland took his fortune to a homemade laboratory in a hut on his estate overlooking the Hudson River in New York, and looked to expand it by developing electric insulators. The universal insulator for electrical coils was shellac, the resinous amber-coloured secretions of the *Laccifer lacca* beetle of southern Asia, where it was mostly used as a wood preservative. By 1900 demand was already outstripping supply (it took 15,000 enthusiastic beetles six months to produce one pound of shellac), and Baekeland believed he might be able to make synthetic shellac at home.

His starting point was something that a dye chemist had made at the bottom of a beaker and then thrown away. Several years before he synthesised indigo, Adolf von Baeyer had been experimenting with phenol (the turpentine-like solvent distilled from coal-tar) and formaldehyde (the disinfectant and embalming fluid obtained from wood alcohol). Thirty years later, Baekeland discovered – after several other Bavarian, Austrian and British chemists had toiled away at different permutations of the phenol–formaldehyde reaction and produced a soft casein plastic known as Galalith or Erinoid – that the residue rejected by Baeyer might form the basis of synthetic shellac, or what he soon called polyoxybenzylmethylenglycolanhydride, and most people called Bakelite.

The key lay in his Bakelizer, an iron boiler that transformed the phenol and formaldehyde reaction under extreme heat from a liquid coating or gummy paste into a hard, translucent and mouldable substance that would set the path for Plexiglas, vinyl and Teflon.*

Bakelite was announced in 1909 with great claims, most of them valid: it was a better insulator than shellac, and it would not burn or boil, fade or discolour. It was not the first plastic (celluloid), but it was the first plastic to be 100 per cent synthetic. Within two decades it encased electrical coils, radio valves and telephone bells, and made a fortune for the company that made the first Bakelite billiard balls. Baekeland won some awards for his work, one of which he particularly treasured: the Perkin Medal.

Among the speakers at the Perkin centenary, Herbert Levinstein

* Almost too good to be true, one pre-Bakelite advance towards synthetic plastic was made when a cat knocked over some formaldehyde into her saucer of milk at the Bavarian laboratory of Adolf Spitteler. This made it curdle into a hard substance resembling celluloid, and Spitteler was then only weeks away from developing an early casein plastic, sometimes known as Lactoid.

told the Chemical Society that Perkin had been 'like a foxhound puppy' dashing here and there in pursuit of knowledge. He thought that this may have been because, like Faraday, he had not attended one of the top boarding schools and thus benefited from an unregimented mind. He said it was wrong for present-day chemistry students to think of him stumbling upon mauve as a lucky break, calling it 'in fact the least accidental of discoveries' on account of his curiosity and systematic methods of inquiry. Levinstein ran through Perkin's career, and closed with the wish that his work would continue to inspire the conquest of tropical disease: 'By biochemical research, organic chemistry may make its greatest contribution to a tortured world.'

A few weeks later, the chemistry scholar F. M. Rowe addressed the Manchester Literary and Philosophical Society. He said there was now a tendency to belittle Perkin because he was an opportunist; as in 1856, pure research still liked to denigrate his commercial instincts. Referring to his death, he revealed that he was unaccountably averse to traditional medicine, calling for his dietician and dismissing his doctor when the symptoms of his pneumonia first appeared. Given what his dyes had achieved since, the irony of this did not go unnoticed.

The anniversary was celebrated with some optimism, for British dye firms had made good use of tariff protection to expand both output and technical innovation. In 1938, ICI controlled about 60 per cent of British sales and had developed important new dyes for the new cellulose fabrics. Before the war, British firms had strengthened their links with IG Farben, predominantly in an attempt to divide up new markets and underwriting fundamental research into new polymers. German occupation of mainland Europe led to the return of old cartels, and a full appreciation of what use the wealth of huge dyeworks could be put.

In 1939, ICI was producing about 200 tons each week of its great new discovery, polythene. Some of this sheathed the new revolution in cable communications. In 1940, the polythene casing of airborne radar enabled the Royal Air Force to detect and intercept German bombers. In Germany, where the war effort initially enjoyed no such privilege, the production of synthetic materials was focusing on rubber and oil.

IG Farben did not officially became part of the Nazi party, but Hitler could not have contemplated war without it.* Initially Hitler had despised the company's internationalism and the employment of the many Jewish chemists that had ensured its prominence (it was classified as a non-Aryan operation), but he realised he needed the synthetic self-sufficiency it made possible. By 1937, the directors and leading scientists of IG Farben who were not Nazis already (or who had not already fled the company) acknowledged that they needed Hitler for future expansion. The combine of dye companies had diversified into strategic raw materials such as synthetic rubber and oil, yet recession and high development costs had almost brought the company to its knees in the early 1930s. But the materials were ideally suited to the modern war machine, and the Nazis placed great value upon them. The depth of this partnership was recorded at Auschwitz.

The grim, resonant phrase 'Arbeit Macht Frei' appeared on posters in the Buna rubber factories of IG Farben before it appeared above the gates of the concentration camps, primarily as a disincentive from joining the trade union movement. Buna (an abbreviation of its main components butadiene and natrium) was one of IG Farben's great synthetic inventions, and, among other uses, had wrapped itself around the wheeltracks of tanks. By the outbreak of war, its two plants were considered insufficient, and in

* Even at the time of the Weimar Republic, Gustav Stresemann, chancellor and foreign minister, noted 'Without I.G. and coal, I can have no foreign policy.' A team of civilian and military experts assigned by General Eisenhower after the war concluded, 'Without I.G.'s immense productive facilities, its far-reaching research, varied technical experience and overall concentration of economic power, Germany would not have been in the position to start its aggressive war in September 1939.'

1941 the search for a new site led the company to establish Monowitz-Buna, the slave labour camp eight kilometres from Auschwitz. About 40,000 prisoners, mostly Jews, were put to work there, and at least 25,000 died. Primo Levi, one of the skilled chemists at the camp, noted in his great testament *If This is a Man* that not one gram of synthetic rubber ever made it beyond the factory walls. One of IG Farben's subsidiaries, Degesch, manufactured Zyklon B, the poisonous gas initially marketed as an insecticide, but swiftly sent to Birkenau in grey crates as the principal method of murdering Europe's Jews.

13
PHYSICAL ACTS

British scientists have created the world's first electronic dictionary of colours, a digital palette containing more than 16 million shades. As a result, colours can now be transmitted electronically and accurately, allowing designers, manufacturers of cosmetics, and fashion workers to exchange images in precisely defined colours for the first time.

The system – developed at the University of Manchester Institute of Science and Technology – has been designated as one of the Government's key industrial projects for the new Millennium and it is already being used by Marks and Spencer and cosmetics manufacturers. You can see what your M&S shirt will look like inside a store, and then out in the street. No more suave mauve creations turning naff pink in daylight.

Robin McKie, *Observer*, December 1998

In 1956, people who liked colour and the effect it had on the world gathered alongside those who worked in textiles, drugs, perfumes and plastics to celebrate 100 years of mauve. People had only good things to say about William Perkin and the change he had created.

By 1956, basic colours were not really the thing any more. New shades tended not to cause a sensation. The big issue was technique – how to put synthetic colours on new acetates and rayons, fabrics that melted when you ironed them. In 1956, the most fashionable colour was black.

The big new synthetic inventions of 1956 found mostly peaceful uses. The Swiss inventor George de Mestral patented Velcro,

Procter and Gamble gave babies disposable nappies, and a Connecticut doctor Vernon Krieble announced the instant superglue resin Loctite. In 1956, a junior advertising executive called Shirley Polykoff coined a winning phrase at the Manhattan agency Foote, Cone & Belding for its client Clairol and the product Miss Clairol. Miss Clairol was the first home-use hair dye that you applied just like a shampoo. It was cheaper and quicker than a professional job, and it brightened up the city streets. Polykoff's slogan was: 'Does she or doesn't she?'

In 1956, Lawrence Herbert joined a small Manhattan company called Pantone, and not long afterwards developed the Pantone global colour language, the world's most widely used colour standardisation and matching system for textiles, cosmetics, paints and inks. And at ICI in 1956, the most significant modern step forward in dye chemistry took the form of reactive dyes, which the company called Procions. These dyes were the first to react chemically with the material, a process involving the addition of sodium chloride and alkali, forming an intense bond with the fibres they coloured, particularly useful in preventing fading on cotton and wool.

In 1956, a year when the United Kingdom produced 90 per cent of the dye used within its shores, the Perkin Centenary began in May on a moody note. An editorial in *The Dyer and Textile Printer* observed that while the world's chemists prepared to whoop it up on an industrial scale, no one in the outside world seemed to care very much. 'It is regrettable, we think, that so little is being done to erect plaques on some of the buildings connected with Perkin. This is a traditional way of commemorating the great and arousing interest in them, and London in particular is full of plaques, many of them to far less important people than Perkin. Yet the London County Council some five years ago declined to place a plaque on his birthplace because of the condition of the building.' *The Dyer* judged this naive; why not a plaque now, and then if it gets knocked down another plaque saying, 'In a house on this site . . .'?

Near the factory, the enthusiasm was not much greater. 'In the Greenford and Sudbury districts, the local authorities have never shown much interest in Perkin (Wembley Borough Council has gone so far as to give the name of Butler, a former mayor, to a garden and recreation ground on the site of the garden in Sudbury in which Perkin grew madder . . .)'*

In the week this observation appeared, Perkin had taken over the halls of England. His reputation was monumental: in the fifty years since the jubilee, in the forty-nine since his death, he had achieved more than he could have dreamed and probably more than he wanted. Almost nothing seemed to exist without him now; guests wore the mauve ties and sat at mauve tablecloths and wore synthetics, and there was dancing to shellac 78s.

The big London events included lectures at the Royal Institution, a soirée in honour of overseas visitors from fourteen countries (at the Worshipful Company of Tallow Chandlers, attended by the President of the Board of Trade), a reception at the rebuilt Guildhall, and an essay competition sponsored by the London Section of the Society of Dyers and Colourists (SDC), for under-26-year-olds, on 'The influence of Perkin's Discovery, and of the synthetic dyes which followed it, on any trade of the candidate's choice'. (Six months later, the London Society reported that the essay competition 'met with a most disappointing response', and only a consolation prize of five guineas was deemed suitable.)

A banquet at the Dorchester was enjoyed by 400. The guest of honour was the Marquess of Salisbury, who felt that insufficient credit had been accorded to Perkin's father, who had risked his savings on the whim of a youth. There was tribute too to Perkin's brother Thomas, who had succumbed to a brain haemorrhage in

* To be fair to Wembley, the Wembley History Society had arranged a small Perkin exhibition at the Barham Park Public Library, and the Sudbury Methodist Church did put up a bronze sign inscribed 'This plaque was placed . . . to mark the centenary of the discovery of the first aniline dye mauve by Sir William Perkin, founder of the original place of worship on this site 1856–1956.'

1891 at the age of sixty. There were telegrams and scrolls from India, Japan, Norway and Pakistan. H. Inouye, the president of the Japan Chemical Association, spoke of his industry's debt to 'Mauve man', and welcomed all the new branches of chemistry he made possible. 'All this, we believe, proclaims the immortal glory of Perkin . . . We take this opportunity of saying that we are very anxious to do all possible in the cause of world peace and the happiness of mankind, working side by side with worldwide chemists and scientists.'

At several events, Perkin's name was used as a tool. He became a symbol of every British pioneer whose work was only fully exploited abroad. There was the mauve story, there was the penicillin story, and many other stories told in the aftermath of empire. There was much truth in them, and they were principally used as a warning, a morality tale. Modern scientists and engineers would say, 'Here's another new development which we mustn't let slip,' and in 1956 Perkin's was the classic example.

In Manchester, the Midland Hotel was host to an event in which the local dyers sported mauve nylon handkerchiefs. The Midlands section of the SDC met for a dinner-dance at the Welbeck Hotel in Nottingham, during which there was a presentation of mauve lace handkerchiefs for the ladies. The Huddersfield section took part in a Perkin item on the BBC's *The Week Ahead*. The Northern Ireland section celebrated at Thompson's Restaurant, Belfast, with an address from the Courtaulds dyer John Boulton. 'I have been driven to reflect upon the question of what kind of man is a Perkin,' he began. 'How comes it that there are so few of them?' He concluded that a Perkin was the sort of man who made great use of the things that happened to him. He was not the sort of man who made anything of the science/industry/art divide, but embraced it all. 'A person of much greater ability as an experimentalist and with a deeper knowledge as a scientist could easily have failed to make Perkin's discovery had he not also something of William Henry Perkin's makc-up . . . he was great because he was what he was, and not because he made the discovery whose centenary we are honouring by our meeting tonight.'

The British Colour Council, which had given Perkin's mauve the classification 225 in its colour index, suggested that there were seven fashionable shades in current use that were based directly on Perkin's first colour: wild orchis, daphne pink, pink clover, sweet lavender, purple lilac, homage purple and violet.

Sir Patrick Linstead, Rector of Imperial College and the chemist who revealed the structure of the important phthalocyanines (a blue-to-green class of fast and brilliant dyes with a metallic core), opened a small month-long exhibition at the Science Museum: samples of natural dyes; a picture of the nineteenth-century woad mill near Wisbech; letters from Perkin, a few of his dyes; a panel on the work of the British Colour Council. A great success, the show was extended by two months.

But not all events were so popular. After the principal celebrations in London, W. Ronald Kirkpatrick, one of Perkin's grandchildren, received a letter from John Nicholls, secretary of the London centenary celebrations, expressing that he 'personally would have liked to see bigger attendances . . . It would seem that inside and outside the industry there appears to be a lack of understanding of the great value to the world [that] colour means in the very joy of living. Anyway, we had quality present on all occasions, if not quantity.'

And the science magazines were full of it. *Nature* suggested that it was quite possible that other chemists would have stumbled upon aniline dyes within a few years had Perkin not, but this in no way belittled Perkin's achievements. The magazine supported Perkin's own claim (in a lecture delivered in 1868) that 'to introduce a new coal-tar colour after mauve was a comparatively simple matter. The difficulties of all the raw materials had been overcome, as well as the obstacles.' Perkin was not just lucky: his initiative, resourcefulness, imagination and determination made him the leading technologist of his day, and Perkin's real qualities and claim to remembrance were 'exactly those which are still needed as Britain faces the social and industrial implications of an age in which coal will no longer form the sole basis of power or of technology'.

In the age of oil, *Nature* detected a growing indifference to the achievements of a coal-tar hero. In the national newspapers, the opportunity which the centenary offered of commending a career in technology to youth was ignored. 'The contrast between the treatment by the Press of the celebrations of 1906 with those of 1956 suggests some neglect of responsibility.' *The Times* almost completely ignored the event, briefly noting the exhibition at the Science Museum. The *Daily Telegraph* had little interest in Perkin or his disciples. But the *Manchester Guardian* made amends with a special supplement, in which it called on various British specialists to demonstrate how Perkin's work had influenced most modern things between 'the vibrant colours of a spring dress and the antibiotic drug saving the life of a desperately sick man'.

It featured Paul F. Spencer, the chief chemist at Cussons Sons and Company, who spoke of how Perkin's synthetic coumarin was widely used in perfumery, how it helped to flavour tobacco, and how a derivative of coumarin had led to a superwhite bleaching agent in the detergent industry. (Coumarin is present as a major constituent in plants such as tonka beans and as a minor constituent in strawberries, cherries and apricots. Following experiments on rats it is now accepted as a carcinogen, and has been withdrawn from many brands of cigarettes.)

But the chief correspondent was Frank L. Rose, Research Manager at ICI Pharmaceuticals, who observed that speculation on what might have been is generally a useless pastime, but there was solid ground for suggesting that 'without Perkin's observation the progress of therapeutic medicine might have been delayed by as much as a generation'. He wrote of the work of Ehrlich and

Domagk, of methylene blue and the sulfa drugs treatment of bacteria, but surmised that had Perkin not been the curious sort, the world might only then (1956) be at the beginning of its understanding of chemistry in relation to disease.

The accompanying display advertisements took a similar line and were probably the first to feature the colour mauve as a dominant motif. One was illustrated by four figures on a blackboard: 328, 238, 434 and 27.6 per cent – co-ordinates in the international CIE system of colour definition, corresponding roughly to the redness, greenness, blueness and lightness of mauve. 'This example indicates the progress made in colour technology in recent times. It is 100 years since Sir William Perkin's discovery. For over 80 of those years Chadwicks of Oldham have been using dyestuffs of better and better fastness . . .'

A familiar name took out an advert summarising Perkin's visit in 1856. 'Not only can Pullars of Perth fairly claim to have played a vital part in launching the greatest discovery in the history of dyeing, but only ten years later they introduced Dry Cleaning . . .' The Shell Chemical Company paid for a line-drawing of a a conical flask placed over a bunsen burner, and next to it a sketch of an oil refinery, the source of its petroleum chemicals today. 'Surely Sir William never dreamed it would grow to this . . .'

The last of the significant celebrations occurred in September at the Waldorf-Astoria in New York. The American Association of Textile Chemists and Colorists and twenty-six related trades gathered for a week of speeches on themes such as 'Color – The Catalyst of Commerce'. In the middle of the week, the fiftieth Perkin Medal was presented to Edgar Britton for his work on the early synthesis of phenol (carbolic acid), whose derivatives are essential in weed-killers, fungicides and other agricultural chemicals. There was a posthumous Perkin Medal for Wallace Carothers for the discovery of nylon. Past recipients included Irving Langmuir for his development of the gas-filled incandescent electric light, Thomas Midgley for his work on non-detonating fuels for internal combustion engines, Robert Williams for the synthesis of vitamin B1 (thiamine), and Charles Hall for the commercial production of aluminium.*

While the centenary celebrants gathered their papers, their partners were treated to a cruise around Manhattan and a show called *Cavalcade of Color*, 'A stage revue with music . . . dramatizing a century's evolution of color as a compelling force in the creation, promotion and popular enjoyment of fashion.' Other rooms at the hotel contained dioramas depicting plastics and pharmaceuticals industries. At nearby Grand Central Terminal, the Eastman Kodak Company staged a continuous slideshow of the preparation and development of modern colour film.

At the hotel, the serious business was conducted by experts in their fields. From food colour to military research, from ophthalmology to the graphic arts – aniline dyes once again had done it all. Some speakers told elegant histories. Morris Leikind, of the Armed Forces Institute of Pathology in Washington, DC, recalled that when the great Dutch physician Hermann Boerhaave died in 1738, he left behind a booklet in which he promised to reveal all the secrets of medicine. It was blank, apart from the instruction 'Keep the head cool, the feet warm and the bowels open.' Only a few of the pages were filled in the next 118 years, Leikind claimed, citing the smallpox vaccination of Edward Jenner of 1798, the discovery of the mammalian ovum by von Baer in 1827, the numerous advances linked to microscopy and the development of anaesthesia in 1846. But after 1856 the blank pages began to fill rapidly. The secrets were 'written in rainbow colours, aniline purple, Bismarck-brown, magenta, methylene blue'.

* The 1999 Perkin Medal was awarded to the Kentucky-born Dr Albert Carr for his work on the discovery of terfenadine (known primarily to hayfever sufferers as Seldane, Teldane or Triludan), the world's first non-sedating anti-histamine, a flagship product for Hoechst Marion Roussel. Dr Carr also developed the anti-psychotic compound M100907, targeting the treatment of schizophrenia, and is named as the inventor on sixty-seven US patents.

Such was the nature of proceedings: moist overstatement in thc cause of tribute. A few guests became a little over-lyrical. One praised Perkin for increasing the colours of war: 'There are the colours produced by flame munitions and weapons; by napalm bomb bursts; the colours of phosphorous shell explosions; the bright colours that mark bomb drops; fuming nitric acid and the gleaming white of the guided missile as it rises majestically and swiftly into the blue atmosphere.'

The most interesting new analysis came from Deane B. Judd at the National Bureau of Standards, Washington, DC, who found that Perkin and his successors had made a significant contribution to the English language. Of the 7,500 colour names identified by this time, over 100 originated directly from synthetic dyes. (That is to say, while almost all the 7,500 could be made artificially, there were over 100 names – including anthracene green and naphthalene yellow – that originated purely from the chemist's workbench. The other sources include 528 flowers (from amaryllis to wisteria), 427 proper names of places (Antwerp brown to Zanzibar brown), 340 pure colour names (black, blue, red), 290 pigments (chrome green), 254 fruits (apricot, banana), 239 foods (brown sugar, yolk yellow), 221 peoples (Tyrian purple, Dutch blue), 214 substances (amber, asphalt), 200 personal names (Robin Hood green, Salome pink), 183 botany (acacia), 149 common things (brick red), 144 natural dyes (indigo, madder), 133 birds (bluejay) and 133 animals (buff – from buffalo). There were 125 jewels (amethyst), 123 metals (brass), 121 geographical elements (glacier blue), 117 alcoholic drinks (absinthe), 107 trees (willow green), 105 atmospherical features (aurora yellow), 83 weather aspects (smog), 82 moods (blue funk), 79 abstract things (triumph blue), 72 romance and passion (golden rapture), 64 minerals (agate), 60 old things (antique brown), 59 end-use (battleship grey), 56 fable and superstition (goblin scarlet), 55 time of day (midnight blue), 50 marine life (coral), 50 undyed textiles (ecru), 46 mythology (Bacchus), 36 ceramic (Wedgwood blue), 31 religious occupations (cardinal purple), and 20 human (nude).

There were hundreds of others. Of the 108 registered names of original synthetic dyes, the two that most people had heard of were magenta and mauve.

In March 1981, a man called Edward G. Jefferson arrived at Manhattan's Plaza Hotel to pay tribute to William Perkin and tell a few stories. He shared the occasion with 500 other chemists, ostensibly to eat duck and award another Perkin Medal to a person of consequence. Jefferson, as president of the American Section of the Society of Chemical Industry, gave the welcoming address.

Everyone knew Jefferson. A few months before, in his role as President of the DuPont chemical company, Jefferson had made a big noise in the business pages when he announced that after more than 60 years, DuPont was getting out of dyes. Since its invention of Lycra, dyes were no longer the most interesting or profitable thing at the company, and it had decided to sell the division.

DuPont had been in dyes since 1917, another venture begun when the war curtailed the supply of dyes from Germany. DuPont's big thing was sulphur black and indigo, but it tried its hand at most colours. It was big on cosmetic and hair dyes.

These days there is hardly a home in the industrialised West which does not contain at least one item trademarked by DuPont – be it nylon, Teflon, Lycra, cellophane or any one of hundreds of medicines and petroleum-based products. But DuPont is not a twentieth-century phenomenon: it is claimed that without DuPont America itself would be unrecognisable.

Éleuthère Irénée du Pont de Nemours, a French nobleman and publisher, met President Thomas Jefferson in Paris in 1784 and helped him draft the peace treaty that ended the American War of Independence. Later, faced with political upheaval, du Pont fled for America with his son, who was persuaded by Jefferson that he should use his knowledge of chemistry to manufacture gunpowder. Du Pont soon had a huge explosives mill, supplied the Union forces during the Civil War, and established an empire that by 1999 had annual revenues approaching $30 billion and employed 84,000 staff in seventy countries.

At the Perkin dinner, the men were still obliged to wear mauve bow-ties (women were now allowed into these events, and some wore mauve ribbons). The dye for the ties had come from DuPont, but it had been an ordeal. 'The DuPont Company was asked to produce a new lot of mauve two years ago, and we confidently assigned this task to our dye experts at the Chambers Works,' Edward Jefferson, a descendant of the president, explained. 'Unfortunately, our first attempt to produce this gorgeous hue failed so badly that the material was flatly rejected by the Society's standards committee and we had to try again. Again we failed, but on the third attempt we were successful, although I'm told that even then our product barely got by. In any event, we were so chastened by the experience that the DuPont Company decided to go out of the manufacture of dyes, and we have just recently sold our entire dye business.'

Roars of laughter from the chemists, who never once believed that Jefferson was being serious. After the meal, the Perkin Medal was awarded to Ralph Landau, from Philadelphia, for important work on nylon and polyester.

Twelve miles west of Landau's home town, out on Route 3, lies Newtown Square, an unremarkable place ignored by the tourist guides. On its perimeter, past several churches and a sign that reads NO GUNNING IN NEWTOWN TOWNSHIP sits an 800-acre wooded estate containing a mansion known locally as the Big House. This is the home of John Éleuthère du Pont, the wealthiest, craziest man in town.

Or at least he was, until he was transferred to jail in 1997, at the age of fifty-eight, for killing an Olympic wrestler at the bottom of his driveway. 'We just thought that the odd things he was doing he was doing for attention,' I was told by Kurt Angle, a former amateur wrestler who has since turned pro. Angle once trained with du Pont on his estate, as part of the Foxcatcher team that du Pont had established in the 1980s. 'He just wanted people to take notice of him and say, "Wow, that du Pont – he can do anything he wants." I don't think anyone ever thought he was going to end up killing somebody.'

The du Ponts long ago lost control of their chemical empire, but the riches remain in the family, and John du Pont's personal fortune has been estimated at $125 million. Like many men too wealthy to work, du Pont has spent many lonely years indulging his passions – shooting, swimming and wrestling – throwing money at things that please him, building up the leading amateur wrestling club in America, building up a private arsenal of handguns and assault rifles.

Examples of du Pont's eccentricity have had them talking in the town for years. He thought there were Nazis in his trees. He told police that he shot the geese in his ponds because they were casting bad spells. A former builder on the estate told of how du Pont believed his mansion was fitted with an oil-spraying device that made things disappear. No one was sure why he shot the wrestler Dave Schultz in January 1996, but at the wrongful-death trial the following year his lawyers spoke of insanity and a 'chemical imbalance'.

In November 1999, David Schultz's widow Nancy settled her claim against du Pont for a sum believed to be $35 million, the largest award resulting from a wrongful-death suit ever paid directly by one person, and $1.5 million greater than the amount awarded against O. J. Simpson.

During his trial for third-degree murder, at which he was found guilty and sentenced to a minimum of thirteen years in jail, the jury heard so much testimony that it took a week for them to agree a verdict. They heard of Dave Schultz's gold medal at the 1984 Olympics, what a great father he was, and how he was a true ambassador for his sport. Schultz had a well manicured beard, and took much pride in his appearance. During the trial, every detail of his wrestling career was examined, including his choice of Lycra leotards, which were described as red – and mauve.

14
FINGERPRINTS

I had lunch with Quentin Crisp the week before he died. We met in the Bowery Bar in Manhattan on the Lower East Side for crab cakes and whisky, and for two hours I sat and gazed in wonder at an old man with mauve hair, the self-styled Stately Homo of England.

Gyles Brandreth, *Sunday Telegraph*, November 1999

The striking portrait of William Perkin by Arthur Cope, once in Perkin's home, found its way to the National Portrait Gallery by Trafalgar Square in 1921. But today it is not on display next to Faraday or Darwin or the other eminent Victorian scientists in the little room set aside for them on the first floor. In the postcard rack in the shop, there is no room for him between Samuel Pepys and Beatrix Potter. The painting is in the basement, the receptionists think. In fact, the archivist locates it in a crate in a store-room south of the Thames, where it lives alongside hundreds of other men and women who are no longer trusted to excite the public imagination.

The bulk of Perkin's life, or at least the material items that have survived it, lie in tissue and plastic envelopes in several large card boxes in another store-room, this one in the lower ground floor of the Museum of Science and Industry in Manchester. The items have been named The Perkin Collection by Peter Crichton Kirkpatrick, one of William Perkin's grandchildren. For several years they were stored at Zeneca Specialties, the dyes and fine chemicals company established by demerger from ICI, but they moved a few miles to the Manchester museum when its archivist died. The collection contains some of the letters quoted here, some formal photographs, a collection of patent certificates, and some of the medals and citations awarded to Perkin on the big anniversaries. Then there are more unusual items: a slab of stained concrete floor from an unspecified Perkin laboratory (probably at Greenford Green); Perkin's science lecture notes from the City of London School, written at the age of thirteen; samples of chemicals, dyed cloth and patent seals; a bow-tie worn at the American jubilee banquet at Delmonico's; and the original 1856 notebook used by Perkin to denote the discovery of mauve. This small book, frayed but still firmly bound, contains the method of preparation of the colour, and some tests upon it. On several pages there are mauve fingerprints.

In the centre of a large glass display case on the first-floor landing of the chemistry block at Imperial College students may read a condensed history of their institution. They learn that in 1873 the Royal College of Chemistry moved to the New Science Schools in South Kensington, now the Henry Cole Wing of the Victoria and Albert Museum. This was called the Huxley building, and conditions were so crowded that some lectures were given at the Albert Hall, where students complained about the noise of the organ. It became part of the Royal College of Science in 1890. And the Royal College of Science became part of Imperial College in 1907. The display case contains a picture of Professor Hofmann but not of Perkin.

Beneath this there is an open science journal, undated but firmly rooted in the heart of Victoria's reign, in which Hofmann presents his 'Remarks on the Importance of Cultivating Experimental Science in a National Point of View'. It begins:

The present century, so rich in discoveries in every department of science, is more especially remarkable for the amount of human activity displayed and the success attained in the improvement of all material interests of society. The rapidity with which we are advancing in this direction is truly astonishing. Every year is fertile in discoveries in science, and almost every day brings forth some new and useful application of it to the purposes of life. Industry is in a state of perpetual advancement. Arts and manufactures which were supposed to have maintained perfection have been entirely superseded by the discovery of new principles and the introduction of new methods founded on them. It may indeed be safely affirmed that in no previous period in the history of the world has every branch of human industry undergone so thorough a revolution as that which has been affected during the last fifty years.

Three floors up, a 60-year-old man called David Phillips works in an office which contains a model of a sailing ship and several large holograms, including one of himself. Professor Phillips is the head of the chemistry department at Imperial, and in the autumn of 1999 he also became its Hofmann Professor, an enviable title which brought no additional wage but a firm bond with history. 'They thought I should have his name,' he explains, 'even though my expertise is not the same as his was. But I was greatly flattered and gladly accepted.'

Professor Phillips is a physical chemist, which means he's interested in quantitative aspects of reaction, kinetics and dynamics. Most of his work involves light and lasers. He teaches several courses each term, and has come to realise that while his students are exceptionally bright and motivated, they don't know much about the history of their subject, which he thinks is a shame. 'Perkin was one of the greats of chemistry,' he reasons. 'Although there is no chair named after him, the college does have a Perkin Laboratory.'

Over the last ten years, Professor Phillips has been involved in the development of dyes for the treatment of cancer. These are dyes injected into the bloodstream of a patient which diffuse throughout the body within a couple of heartbeats. They are then lost through the normal mechanisms – the liver and spleen – and are excreted.

But the dyes used are selected because they are selectively retained in tumour tissue. If you wait a couple of days after the injection, there is more dye in a tumour than there will be in the normal surrounding tissue. So then you blast the area with a high-powered laser and, if all goes to plan, destroy the tumour.

The dyes are called phthalocyanines, big plate-like molecules used predominantly as pigments and food dyes. In the 1970s there was a lot of work synthesising the structure of these molecules, experimenting with the introduction of new substituents around the rings.

The dyes are based on molecules that go back to 1913, when a man called Meyer-Betz began experimenting with pigments called porphyrins, widespread throughout nature and in haemoglobin. He injected himself with haematoporphyrins, and then, for reasons best known to himself, went out in sunlight and suffered severe burns (the porphyrins were a sensitising agent in his skin, absorbed sunlight, and fried him). Not much was made of this until it was realised that this action might be applied to cancer treatment by aiming high intensity light into a diseased organ. And it was not until the emergence of very intense diode lasers in the mid-1990s that the beam could be focused down accurately into fibres and inserted into the body.

Professor Phillips strayed into this area when he was working for the detergent company Unilever. The company was interested in developing a cold-water washing powder, a bleaching formulation that would work in the developing world.

'One of the things that occurred to us was that you could use sunlight,' Professor Phillips remembers, 'so we began looking for an additive [a colour-free dye] that became active in sunlight and destroyed the stains. The dye absorbs light, becomes excited and therefore has a lot of excess energy, and the excess energy is transferred to oxygen, which is in the water or the fabric, and it creates an excited state of oxygen called singlet oxygen. This then attacks the chemicals which are in the stain.'

But Unilever was beaten to the market by Procter and Gamble, so Phillips and his team stopped doing that work. 'But at the time I met a medic who alerted me to the fact that a dye had been announced that had been used for this photo-dynamic therapy of tumours. We wondered, "Wouldn't it be interesting if our bleaching dye actually had this effect as well?" We got a project going, and it's been one of those astonishing things that have been successful at every turn.'

Tens of thousands of patients have been treated in this way. The treatment is designed for primary tumours, and as yet is unable to treat rogue cells throughout the body. It is very widely used in ear, nose and throat cancers, and has saved a great amount of brutally disfiguring surgery. When Professor Phillips lectures on this subject he shows his audience some slides of a woman who developed a melanoma on her nose, a patient who would normally require surgery to excavate the bridge of her nose and a large part of her cheek; probably she would have lost an eye as well. The new treatment, however, takes half an hour, and after two weeks you cannot find a scar.

Photo-dynamic therapy is very widely used in colon cancer, and has had some success in surface brain tumours – as a cleaning-up process after surgery. The most exciting uses are in pancreatic tumours, where surgery is often considered hopeless. There has also been some success with prostrate and bladder cancers, and with breast tumours.

'It's been very satisfying,' Professor Phillips says. He calls his dye work 'a novel application of a casual observation', one of the frequent cases of serendipity in his line of business. 'I suppose you could draw a parallel with Perkin, who was looking for a medical application and found a dye, while we were working with dyes and found a medical application.'

Recently the emphasis has switched to examining whether a

number of other, microbial, diseases might also be susceptible to the same treatment. 'We've shown in a large number of cases involving bacteria and fungi and yeasts that you can deactivate these beasts by adding the dye, irradiate with red light, and you knock these fellows out. We've been doing work with a dental hospital, knocking out oral microbes, and knocking out gut microbes in the lab. And you can destroy a lot of viruses this way – including HIV.*

'Chemistry now is a very dynamic subject,' Professor Phillips believes. 'It used to be said, up until about 1986, that chemistry was not exactly dead, but it was felt that we were mopping up. It was felt that there weren't going to be any spectacular new breakthroughs, and we understood it all; we were filling in bits of knowledge, and then applying it. But then Harry Kroto discovered Buckminsterfullerene [a molecule in which 60 carbon atoms resemble the structure of a geodesic dome], and there's a whole new family of carbon-based compounds which we didn't know existed. So chemistry is far from dead.'

In 1999 there were six new companies established on the back of research work at Imperial. 'It's a great trend,' Professor Phillips says. 'When I arrived here, there were still professors who would throw their hands up in horror if you were too involved in industry. There was this feeling that people who were very good at science shouldn't dirty their hands. It was believed that that path was only for the second-raters.'

In April 1944, eighty-eight years after Perkin's first failure, Robert

* Photo-dynamic therapy has also found uses in the treatment of age-related macular degeneration (AMG), the leading cause of sight-loss in the elderly in the developed world. AMG is caused by blood vessels leaking into the central part of the retina, causing hundreds of thousands of cases of blindness and several million cases of reduced vision each year. The therapy uses the light-sensitive dye Visudyne. Once this has been distributed around the body, a beam is shone through a lens into the eye, and the leaking blood vessels are destroyed.

Woodward, probably the leading American synthesiser of his generation, finally discovered how to make quinine. It was still the most effective treatment for malaria.

Its formula – $C_{20}H_{24}N_2O_2$ – had been identified in 1908 by Paul Rabe at the University of Jena, but only successful synthesis confirmed this formula. The need had never been as pressing as during the Second World War. The Japanese occupied the Dutch East Indies and cut off the main source of quinine, and American troops were suffering. As it turned out, Woodward's work (and that of his post-doctoral student William von Eggers Doering) proved too complex to be commercially viable – it needed fifteen steps just to reach quinotoxin, one of quinine's main constituents and the starting point of synthesis. But their work contributed to the formulation of other treatments, and suggested that the eradication of this most persistent of killers was but years away.

The cause of malarial transmission had been identified in 1897, with the aid of synthetic dyes. In Secunderabad, India, the British poet and public health worker Ronald Ross located the malarial parasite in the body of a mosquito that had previously fed on an infected patient. Malaria was found to be caused by four species of protozoan parasites, the most lethal being *Plasmodium falciparum*. The parasites are transmitted by the saliva of female Anopheles mosquitoes, found primarily in sub-Saharan Africa, and are able to kill within a day of the onset of symptoms.

The first recorded case of a cure by a quinine substitute fell, by accident, to Paul Ehrlich, who found that the methylene blue dye he had used to locate the malaria in a German sailor also eradicated it. But there was no effective laboratory trial to analyse this treatment, and the true era of synthetic anti-malarials began only after the identification of avian malaria (thus enabling controlled tests), and after the disease had ravaged Western armies. The breakthrough came with mepacrine in 1930, a German discovery made after some 14,000 different compounds had been tested to resemble substances close to both methylene blue and the molecular structure of quinine identified by Rabe. The drug was valuable to all sides in the Second World War: chemists at ICI cracked the formula by 1938 and made in the region of 2,000 million tablets annually for soldiers in the Far East. But there were many side effects: mepacrine betrayed its origin as a dyestuff by its yellow colour, and it tended to turn the skin of its users the same shade. New variants soon followed – chloroquine, nivaquine, proguanil and mefloquine (brand name Lariam) – but these too caused side-effects, and in time malaria evolved resistance to them all. In the late 1990s, a new anti-bacterial drug called fosmidomycin was being developed in Germany with promising early results in mice. By analysing the genome of *Plasmodium falciparum*, a key enzyme may be inhibited; the research is being conducted at Justus Liebig University, Giessen.

Why is this still of concern? There are 300 to 500 million new cases of malaria each year, of which 1.5 to 2.7 million will result in death, most of them children under the age of five. Clearly the control of the disease has been only partially successful, although it is often regarded in the West as yesterday's complaint.

Several large pharmaceuticals companies face accusations that they care little for most victims of malaria because these people could never afford to pay for the end-product of successful research. But there is much growing evidence that malaria is once again on the rise in Europe and in other areas where it was once considered eradicated. Some blame the increase on the banning of the toxic and environmentally destructive pesticide DDT in the 1970s.

At the beginning of 2000 there were a number of anti-malarial vaccine trials in progress, and one of the most promising emerged at the Naval Medical Research Center at Rockville, Maryland. As in Giessen, the project involves DNA, specifically the DNA of the malaria parasite *Plasmodium falciparum,* injected into humans. The programme is directed by Captain Steve Hoffman, a specialist in tropical medicine for almost twenty years and a long-standing member of the World Health Organisation's malaria steering committees. The DNA that he administers is incorporated into the body's cells. It enters the nucleus, where it is transcribed by the human cell into RNA, which is then transported out of the nucleus and translated into protein. The body then recognises this protein as foreign material, and mounts an immune response against it.

Malaria may just be the beginning; if the DNA vaccine works in principle it may herald similar vaccines against HIV, hepatitis C and other infections. Dr Hoffman's first proof of protection (in mice) was published in 1994. The first human trials (in twenty volunteers recruited by posters and newspaper advertisements) were published in 1998. A far more complex clinical trial began in January 2000, and three more are planned.

Artificial dyes are employed every step of the way – in the analysis of the *Plasmodium* chromosome, in blood tests, in the examination of the vaccine results. 'We use Wright-Giemsa stain for blood smears,' Dr Hoffman says. 'We have also used acridine orange for staining and diagnosing malaria in the past.'

Dr Hoffman liked the symmetry his work created with Perkin's struggle for quinine. Slowly, one Victorian adventure was reaching its intended conclusion.

Of all the contemporary uses for synthetic dyes, none would have bemused William Perkin as much as their employment in the detection of crime and infidelity.

In October 1999 the first international conference on forensic human identification took place over three days at the Queen Elizabeth II Conference Centre in Westminster. People leapt up to the podium to address the issues of dye work in scene-of-crime ball-of-foot evidence, the fluctuating value of personal ear-prints, and the DNA identification of the victims of Swissair Flight 111.

In the exhibition hall, Scott Higgins, European product manager for PE Applied Biosystems, was marketing his human identification products. PE Applied Biosystems is the life sciences division of the Perkin-Elmer Corporation, a company well versed in genetic analysis, molecular diagnostics and microbial identification. PE's science plays an important role in gene discovery and genetic disease research; the other division of the PE Corporation is Celera Genomics, the private company responsible for the deciphering of the human genetic code.

The basis of the technology is based on dRhodamine gel dyes, a method of separating both proteins and DNA. The company has developed a technology of running fluorescently labelled DNA patterns past a detector, providing a different-coloured peak for each base. As Higgins explained, 'You had your G, your A, your T and your C (the basic codes of DNA), and each would have a different colour. Red is a T, the black should actually be yellow but it doesn't print out so well, the A is green, and the C is blue.'

This analysis is used in criminal intelligence databases for generating and storing profiles of suspects. It is also used to analyse stains from crime scenes – any biological substance – blood, semen, hair. The sensitivity of the technology enables it to generate a profile from flakes of skin. It is frequently used in analyses of disaster scenes, and is increasingly employed as a security device. 'In the United States wealthy people take mouthwash samples of a child,' Higgins explains. 'And then if there is a kidnapping that child can be easily identified.'

A less complex forensic home-testing kit with a similar aim has been selling well in Tokyo. Here, wives who suspect their partners of cheating have been able to buy two chemical aerosol sprays that detect the presence of semen in underwear. You use the sprays on the garment one after the other. If fresh semen is present, it will turn bright green.

The precise formula of the sprays is a secret known only to the Tokyo Gull Detective Agency, but it is believed to be similar to the Acid Phosphatase Test widely used by forensic scientists in modern police departments investigating cases of sexual assault. In this, alpha-naphthyl phosphate is sprayed onto a sample, and any semen present will react to produce alpha-naphthol. The application of a second chemical, often a diazyl dye, will produce a vibrant purple.

In Germany in November 1999, Bayer and Hoechst, jointly known as Dystar, announced plans to merge its dye operation with BASF, thus creating a company of 4,700 employees and annual sales of about £720 million, almost a quarter of the annual world dye market of about £3 billion. Approximately 30 per cent of its output was expected to be sold in Asia, while 40 per cent would be sold within Europe. At the time of this announcement, a name for the new dye company had yet to be agreed.

IG Farben in Liquidation, the official name of the company which still operated more than fifty years after its parent company was disbanded by the Allies (and ostensibly still exists in order to dissolve itself), still has more than 200 shareholders. It paid DM30 million of compensation to the Jewish Claims Conference in 1957, but at the beginning of 2000 was still involved in disputes over further individual payments to its former slave workers.

•

On St Valentine's Day 2000, the public relations officer at Yorkshire Chemicals PLC in Leeds had some romantic news to mark the company's centenary year – or as romantic as things ever got in the modern dye trade. Penny Netherwood said the company was to have a new corporate colour: mauve. She pronounced it 'morv', like the Victorians.

Yorkshire Chemicals is the fourth largest textile dye works in the world, just a few hundreds of thousands of kilos behind the Bayer/Hoechst/BASF alliance and Ciba and Clariant of Switzerland. In November 1999 Yorkshire had engineered a reverse takeover of the dyestuffs division of the American company C. K. Witco, and in so doing expanded its product range from acid dyes for wool and nylon fibres and cationic dyes for acrylic fibre, to one which also included reactive dyes for cotton. It now had global sales of £170 million, still only one quarter of its German rival, still only 5 per cent of the total market, but a serious contender again for the first time in many years.

The company formed in 1900 from a merger of eleven firms engaged primarily in natural dyes, a company seemingly untouched by the advances of the previous fifty years: indigo still came from India and Madagascar, logwood from the Caribbean, camwood from West Africa and cochineal from Tenerife. On one site men 'as yellow as canaries' ground turmeric to colour piccalilli. The daughter of one of the founders – Annie – married a son of William Perkin, Arthur George.

These days the machinery at the Hunslet Road and Kirkstall Road sites produce the latest artificial colour for the latest synthetic fibres, but Yorkshire Chemicals is facing difficult times despite its recent acquisition. Over-capacity. Tumbling prices. Huge safety-testing costs. Impossible competition from the Far East.

'When I first came into this industry it was incredibly exciting,' says John Shaw, the business development director at Yorkshire. 'I used to show clients new colours every few weeks, and say, "Isn't this wonderful?"'

Shaw says that when he started in the 1960s, the job was all about technical support, innovation and helping people solve problems. The last thing he discussed as a salesman was price, but now this is everything.

Shaw, who is fifty-five and comes from Wigan in Lancashire, worked in dyes for ICI for thirty years, leaving not long after ICI (Zeneca) sold its textile dye division to BASF in the summer of 1996. The sale to the Germans told you what sort of industry this had become, he says – a desperately competitive one. The market demands that the colour from his factory now comes in granulated form, like freeze-dried coffee. Easier to measure, and to dissolve.

The first problems came from India, and then Japan and Taiwan and China, the cheap imports that arose from a mixture of low-cost labour, pirated technology and some genuine innovation. Almost all the major European companies established partnerships in India and the Far East, but soon found their expertise duplicated by their competitors. Shaw has the figures in his head, but he struggles to believe them: in India there are more than 600 manufacturers of reactive dyes, 'one under every railway arch'; in 1993 the Chinese exported about 5,400 tons of dyes, but by the end of 1998 the figure was almost 60,000 tons; in 1989 the world price for synthetic indigo was $22–$24 per kilo, but now it's coming out of China at $6 a kilo. Then there are stringent environmental controls and the testing fees, the strict health regulations that ensure each new molecule of colour costs between £100,000 and £250,000 to approve, not including perhaps £200,000 already spent on molecular design.

Against this background, Yorkshire Chemicals' recent takeover seems more a bid for survival than a triumph of expansion. Shaw thinks about what happened at Ellesmere Port. Once a German plant, then British after the war, the site enjoyed a huge boom in the 1960s and 1970s supplying the demand for blue jeans. In the 1990s it was back under German ownership, and was closed down for good in July 1999.

'The same pots and pans that make dyes make pharmaceuticals,' Shaw reasons. 'If you discover a new chemical that solves a basic health problem, then the pharmaceutical industry will cover all its costs and potentially make a huge profit. That used to be the case in the dyes business – the invention of a new dye would command a premium that would reward the cost of invention and bringing it to market. Sadly, that's no longer the case.'

At 425 Oldfield Lane, Greenford, the pub menu at the Black Horse contains a little local history. The pub overlooks the Paddington arm of the Grand Union Canal, opened in 1797 to link Limehouse Docks to Birmingham. 'It was sensibly decided to place a hostelry every two hours, offering refreshment to both bargees and their horses.'

The canal is now all houseboats and lazy tourist trade. It is dirty and fat with waste, and no longer the colour of wine.

In the pub there is no memory of Perkin, no sepia artefact or silk sample, and his colour factory opposite has long gone. The pub is now popular with the employees of the new businesses on Perkin's land – the Hovis distribution point for British Bakeries and the GlaxoWellcome company.

Glaxo is proud of a nineteenth-century blouse in its possession dyed with mauve produced at Greenford Green. The company showed it off at the centenary celebrations, and the editor of its in-house magazine made topographical and scientific connections between then and now. Some of it was merely tenuous marketing: Perkin set up his works in 1857, just as Joseph Nathan was establishing a business in Wellington, New Zealand that would later become Glaxo Laboratories; Perkin utilised coal-tar waste to find colour, while Glaxo used a sisal waste product to synthesise cortisone; Perkin was moved to study chemistry by observing the formation of 'beautiful crystals', and did not some beautiful ruby red crystals isolated by a pharmaceutical giant many years later yield forth vitamin B12?

In 1956, Glaxo employees claimed to 'own and tread daily upon the ground that Perkin once owned and trod'. Managers lunched at The Cottage, the home of Perkin's nephew Alfred, the place where Perkin would entertain overseas visitors who had come to buy his dyes. 'Being himself a teetotaller, Perkin would not refresh his visitors at an inn, even though the Black Horse stood close by . . .'

The pub has survived, but The Cottage, where William Perkin once had his laboratory, has disappeared, to make way for more drugs and a car park.

*

A mile away, on Roxeth Hill, down the road from Harrow School, two large green signs proclaim: 'The Millennium is Christ's 2,000th birthday. Worship Him here – now.'

This is Christ Church, designed by Sir George Gilbert Scott a few years before he modelled the Albert Memorial, and it is where William Perkin is buried. But the search for the grave is a tricky one; many tombstones, most from the nineteenth century, are fading fast; some, separated from their plots, are piled up against walls.

Ralph Goldenberg, the vicar here, lives in a house on the edge of the graveyard. He said he had no idea who Perkin was, or where he was. He explained he had only been in Roxeth for eighteen months, and said that the church administrator would be around to answer any enquiries in three days.

But when Fran Caldecourt examined the burial records she found that Perkin's name was absent.

'How strange,' she said.

His name did, however, appear in a survey of the burial ground conducted in 1991. This revealed that Sir William had since been joined in his grave by Frederick Mollwo Perkin, his son (in 1928), Alexandrine Caroline Perkin, his second wife (1929) and Sacha Emilie Perkin, their eldest daughter (1949) – stacked on top of each other like pancakes. The survey disclosed that it was marked by a large arched headstone with white marble kerbs and chippings, and at the base lay an inscription from Revelation: 'Blessed are the dead which die in the Lord from henceforth'.

One afternoon at the beginning of February 2000, Fran Caldecourt walked around the burial ground in search of Perkin, up and down the narrow muddy lanes with a copy of this survey and a photocopied site-plan of the crumbling plots of forgotten Victorians.

'He should be just here,' she said, 'adjacent with the end of the church.' But she could find no trace of him.

AUTHOR'S NOTE

After all these years, everyone has their own mauve.

In the duty-free catalogue on Virgin Atlantic there is a picture of Eddie Izzard with his usual painted fingernails, and the option to buy an item called Virgin Vie Spring Nail Polish. 'Make your friends green with envy, and your nails glossy green or blue or mauve.' In an issue of *Talk* magazine there is an item about how the Old Vic theatre in London, once home to towering performances by John Gielgud and Laurence Olivier, had redecorated its dressing rooms. Vivienne Westwood, Stella McCartney and Tommy Hilfiger enjoyed adding shine and irony to the fusty old rooms, and the theatre's chief executive Sally Greene was proud of their work. Commenting on McCartney's efforts with many mirrors, she said, 'It's for the times when you're feeling fabulous.' In comparison, an ante-room done up in sombre Rothko mauves was 'for when you've just died'.

In movies, the word has been used imaginatively. In Bruce Robinson's screenplay for *Withnail And I*, the lecherous Uncle Monty defines Withnail with the disparaging phrase, 'He's so mauve.'

There is no comfort in books. The word has moved on since Thomas Beer defined American life in the 1890s as the Mauve Decade and Tom Wolfe chose Mauve Gloves & Madmen as his fictional firm of caterers in the 1970s. In a story collection by Will Self called *Tough, Tough Toys for Tough, Tough Boys*, some of the toughest boys were drug dealers, and they spent some time in a suite at the Ritz. The flock wallpaper was purple, the books on the shelf had been bought second-hand by the yard, and the carpet was – predictably by this time – mauve.

I imagined there was some sort of dare being played out, a bet placed at a writers' gathering, whereby all those who had anted up agreed to place the word mauve in the first chapter of their novels. In the opening chapter of John Lanchester's *The Debt to Pleasure*, the narrator takes us on a tour of his brother's second-division boarding school, and describes the portraits of headmasters hanging in the refectory. One of them 'suggested either that the artist was a tragicomically inept doctrinaire cubist, or that Mr R. B. Fenner-Crossway MA was in reality a dyspeptic pattern of mauve rhomboids'.

In the opening chapter of *England, England* by Julian Barnes, the narrator talks of childhood memories and the delights of a jigsaw puzzle of the English counties: '. . . they would try to recall the colours of the pieces . . . had Cornwall been mauve, and Yorkshire yellow, and Nottinghamshire brown, or was it Norfolk that was yellow – unless it was its sister, Suffolk?'

In the first paragraph of *The Witch of Exmoor* by Margaret Drabble we learn that 'The windows are open on to the terrace and the lawn, and drooping bunches of wisteria deepen from a washed mauve pink to purple. The roses are in bloom.' This is the most common mauve these days, the mauve of gardens. But even here, horticulturists are as subjective as novelists: even now – particularly now – mauve isn't one colour, the way orange tends to be. Mauve moves from lilac to lush purple depending on the light, on descriptive prowess, on memory, on upbringing. Everyone has their own mauve, whatever it may be.

Very near the beginning of *The Way I Found Her* by Rose Tremain, a character also recalls a gentle childhood. 'It was the day before we left our house in Devon and went to Paris. I could describe it as the day before my real life began. Mum was wearing a little mauve skimpy top, and a drapey kind of shirt she'd bought from an Indian shop.'

In this way, mauve is often a remembered colour, a shade from one's past. It isn't often a high-tech colour. But early in 1999 a man called Geoffrey Hughes, an Englishman living in San Francisco with a weird ambition, had given the shade a new twist. Hughes hung out at an aircraft hangar in the Mojave Desert where his company, Rotary Rocket, had erected something called the Roton. This was intended to be the world's first multi-use private space rocket, something that would send satellites up to the beginning of space and then return for another payload a few hours later. Eventually the rocket would also carry wealthy passengers who would pay to play astronauts.

The prototype was impressive, but visitors were also struck by the huge mauve curtain swaddling the base of the rocket engines. 'The curtain is somewhat in our house colours,' Hughes explained. 'The same as the girders in the hangar. The curtain is to hide the rather grotty underneath of the 'air stairs' that are used to board the vehicle and hide the computers that are underneath. The curtain also acts as a modesty panel for people climbing the stairs with skirts on.'

Perkin would have liked that: his very own colour helping women and Scotsmen travel towards the moon.

Predictably, I dreamt about Perkin a few times. On one occasion he left me trivial information: his fellow workers at Greenford used to call him Stainchild; how awful Mrs Swaffield was on harmonium; how the ties at Delmonico's were the wrong shade; and how glad he was that Koch and Ross had received the Nobel Prize. He said how much he missed his father, and how he hoped they'd master artificial quinine soon. He said he wished to be remembered for his work on magnetic rotary power.

On another occasion he told me of the terrible state of his grave. The madder he planted had come through his coffin to push him sideways. There was a strange formation of gases. Before his burial his family had covered his body with silk squares, each a different colour of his own making. The magenta on his hands had caused great eruptions, made worse by leakages from the canal near his factory – effluent that had once burnt the grass. And coal was now bursting up into his back and sides, sharp rocks tangled with logwood, safflower, sandalwood and dava. So beleaguered, he said he felt like one long bruise.

On my visits to Imperial College it was apparent how slender was the connection between the current batch of ambitious chemistry students and their Victorian predecessors. The future was too exciting to look back. As they entered the Perkin Laboratory for their latest analysis of complex carbons, the objects around them didn't seem to be the subject of particular concern. But they wrote their theses on blueberry iMacs, photographed their friends on one-use Kodaks, sprayed themselves with Calvin Klein and masked their headaches with aspirin. Their clothes were not all black; some of them wore blue, green, even yellow. And they walked around as if these colours were the most natural thing in the world.

ACKNOWLEDGEMENTS

I am extremely grateful to everyone who helped me with this book, particularly those who gave up their time to be interviewed and share their thoughts. In addition to those who appear in the text, I am indebted to Tony Travis, of the Sidney M. Edelstein Center at the Hebrew University of Jerusalem, for his scholarly assistance and personal advice. The local history provided by David Leaback has also been very informative.

Many of the archival documents relating to William Perkin are the property of his great, great grandchildren Helen Beaufoy and Michael Kirkpatrick, and I would like to thank them for their encouragement with my research and the supply of photographs.

I owe a debt also to Penny Feltham and Jean Horsfall at the Museum of Science & Industry in Manchester, where the Perkin documents (the Kirkpatrick Collection) are housed, and to Robert Bud for his suggestions and contacts, Leo Slater and colleagues at the Beckman Center for the History of Chemistry in Philadelphia, Penny Netherwood at Yorkshire Chemicals, Anne Barrett of the Imperial College archive department, Sarah Burge at the Colour Museum, Bradford, and the helpful staffs of the Bodeleian Library, Oxford, the London Library, Imperial College Library and the Radcliffe Science Library, Cambridge.

Professor Bill Griffith at Imperial and Andrew Bud read the manuscript and made very helpful suggestions.

This book would also have been the poorer without Brad and Joy Auerbach, Spencer Pack, William Brock, Luke Vinten, Ted Benfey and Roald Hoffmann.

Pat Kavanagh, Rosemary Scoular and Vanessa Kearns at PFD offered their normal level of skilled counsel and enthusiasm. My editor, Julian Loose, has been unfailingly inspiring and supportive, and I am also grateful to his assistant Carrie O'Grady for her perceptive suggestions and photo research .

The recipe for the microscale synthesis of mauve is adapted, with kind permission, from the work of Rhonda L. Scaccia, David Coughlin and David W. Ball, Department of Chemistry, Cleveland State University, published in *Journal of Chemical Education*, Volume 75, No 6, June 1998.

In addition to the documents listed in the bibliography, I have referred to a number of newspaper articles, predominantly from the time of the jubilee and the centenary of the discovery of mauve (1856 and 1906). Recent issues of the *Journal of the Society of Dyers and Colourists* and *New Scientist* have also been useful.

BIBLIOGRAPHY

Abbreviations

BJHS *British Journal for the History of Science*
JSA *Journal of the Society of Arts*
JSDC *Journal of the Society of Dyers and Colourists*

Documents and Journals

From the Kirkpatrick Collection at the Museum of Science & Industry in Manchester:
Letter from A. W. Hofmann to Perkin, Royal College of Chemistry, 23 October 1856
Letters from R. Pullar to Perkin, Mill Street Dye Works, Perth, 12 June 1856, 14 January 1857
Letter from J. Pullar to Perkin, Keirfield, Bridge of Allan, 14 May 1857
Letter from L. Pasteur to Perkin, Ecole Normal Supérieure, 5 September 1860
Letters from H. Caro to Perkin, Mannheim, 10 December 1881, 29 May 1891, 28 June 1906; Freiburg, 31 May 1885
Letter from H. Caro to Lady Perkin, Mannheim, 19 January 1908
Patent specification No 1984, 26 August 1856
Letters from W. H. Perkin Jnr to Perkin, Fair View, Fallowfield, 21 February and 10 March 1906
Draft speech by C. Liebermann, Royal Institution, 26 July 1906

Souvenir of Perkin's visit to America, containing newspaper cutting and photographs
Letter from J. W. Bruhl to Lady Perkin, 20 November 1907
Perkin's early science lecture notes at City of London School, 1851–2
Perkin's laboratory notebook, 1890–93
Perkin's notes written on RMS *Umbria*, September 1906
Letter from Perkin to H. Caro, The Chestnuts, Sudbury, 25 May 1891
Hymns for Perkin's funeral service, 21 July 1907
Chemical Society Appeal for the Perkin Memorial Fund
Perkin and T. D. Perkin reply to the Bill of Complaint of Brooke, Simpson and Spiller, 2 December 1874
Press cutting relating to jubilee celebrations and various obituaries of Perkin and Lady Perkin, 1906 and 1907
Perkin's own copy of the official book commemorating the Jubilee, 1906
Perkin's notebook recording the discovery of mauve, 1856, courtesy of the City of London School
Perkin family trees, undated
Chemical Society booklet on the life and work of W. H. Perkin Jnr
Various scrolls, illuminated tributes, honorary degrees and medals awarded to Perkin in 1906′

Barrett, Anne, *Perkin's Green*, The Glaxo Volume, pp. 43-47, 1956
Benfey, Theodor, 'The Editor's Safari: Jerusalem's Edelstein Centre and Three Anniversaries', Beckman Centre News, Summer 1992
Bentley, Jonathan, 'The Work in England of A. W. von Hofmann, Professor of Chemistry at the Royal College of Chemistry 1845–65', Thesis for the Degree of Master of Science in the History of Science, University of Leicester, 1969
Berrie, John, *Specimens Illustrating New Methods of Dyeing and Finishing Silk*, Great Exhibition Catalogue, 1862
Boulton, John, 'William Henry Perkin', JSDC, Vol. 73, No. 3, March 1957, pp. 81–5
Brightman, R., 'Perkin and the Dyestuffs Industry in Britain', *Nature*, Vol. 177, May 5, 1956

Cliffe, W. H., 'In the Footsteps of Perkin', JSDC, Vol. 72, No. 12, 1956, pp. 563–6
– 'The Dyemaking Works of Perkin & Sons: Some Hitherto Unrecorded Details', JSDC, Vol. 73, No. 7, July 1957, pp. 312–28
Cronshaw, C. J. T., *Through Chemisty, Adornment*, The Royal Institute of Chemistry of Great Britain and Ireland, London, 1949
Crookes, William, 'Chemical Products: The Application of Waste', *Popular Science Review*, London, 1863
Dawson, Dan, *On Azo Colouring Matters*, The Society of Dyers and Colourists, 1884, pp. 12–19
Dickens, Charles, ed., 'Perkins' Purple', *All the Year Round*, London, 30 April 1859
Finley, Alexander, and Mills, William Hobson, 'William Henry Perkin, Jnr', in *British Chemists*, London Chemical Society, 1947
Gilbert, Kerry, 'The Cultivation of Woad (*Isatis Tinctoria*): Agronomy, Physiology and Biochemical Aspects', PhD, University of Bristol, 1997
Homberg, Ernst, 'The Emergence of Research Laboratories in the Dyestuffs Industry, 1870-1900', BJHS, 25, 1992, pp. 91–111
Hooper, C. J. W., 'Celebration of the Centenary of the Discovery of Mauve by W. H. Perkin', JSDC, Vol. 72, No. 12, December 1956, pp. 566–73
Hornix, Willem J., 'From Process to Plant: Innovation in the Early Artificial Dye Industry', BJHS, Vol. 25, 1992, pp. 65–90
Hummel, J. J., 'Notes on the Application of Alizarine and Allied Colouring Matters to Wool Dyeing', JSDC, November 1884
Hurst, George H., 'On The Use of Coal-Tar Colours in the Manufacture of Pigments for Painters, etc.', JSDC, 25 Febrary 1890.
Illustrated London News, 2 and 30 January, 6 February, 6 April 1858
Johnson, A., and Turner, H. A., 'Synthetic Dyes from the Time of Perkin', *The Dyer and Textile Printer*, 11 May 1956
Kauffman, George B., 'Pittacal: The First Synthetic Dyestuff', *Journal of Chemical Education*, Vol. 54, No. 12, December 1977
Leaback, David, 'What Hofmann Left Behind', *Chemistry and Industry*, 18 May 1992

– 'Chemical enterprise from "The Elephant"', *Chemists in Britain*, April 1992, pp. 340–43
– 'Perkin's Pioneering Enterprise', *Chemistry in Britain*, 24, No. 8, August 1988
– 'Discovery in the East End, *East End Record*, 12, 1989, pp. 2–16
Levinstein, Dr Herbert, 'Perkin's Adventure and What Has Become of It', *The Dyer and Textile Printer*, December 1938, pp. 495–6
Levinstein, I, 'On the Coal-Tar Colour Industry', *The Dyer and Calico Printer*, 15 October 1890
Meldola, R., 'Obituary Notices of Fellows Deceased: William Henry Perkin', *Proceedings of the Royal Society of London*, 80, June 1908
Meth-Cohn, Otto, and Smith, Mandy, 'What Did W. H. Perkin Actually Make When He Oxidised Aniline to Obtain Mauveine?, *Journal of the Chemical Society*, Perkin Transactions 1, 1994, pp. 5–7
Meth-Cohn, Otto, and Travis, Anthony S., 'The Mauveine Mystery', *Chemistry in Britain*, July 1995, pp. 547–9
Morris, Laurence E., 'The Genius of Perkin', *Dyer, Textile Printer and Finisher*, Vol 115, 11 May 1956, pp. 747–64
Morris, Peter J. T., and Travis, Anthony S., 'A History of The International Dyestuff Industry', *American Dyestuff Reporter*, November 1992
Nieto-Galan, Agusti, 'Calico and Chemical Knowledge in Lancashire in the Early Nineteenth Century: The Life and Colours of John Mercer', *Annals of Science*, 54, 1997, pp. 1–28
Peel, R. A., 'Perkin and the Scottish Alizarin Dyers', *The Dyer and Textile Printer*, May 1956
Perkin, Arthur George, 'Constituents of Natural Indigo', *Journal of the Chemical Society*, 91, 1907, pp. 435–40
Perkin, Arthur George, and Bloxam, W. Popplewell, 'Some Constituents of Natural Indigo', *Journal of the Chemical Society*, 91, 1907, pp. 281–8
Perkin, F. Mollwo, 'The Artificial Colour Industry and Its Position in this Country', JSDC, December 1914
Perkin, William. H, Cantor Lectures: 'On the Aniline or Coal Tar Colours', JSA, 1 January 1869, pp. 99–105; 8 January 1869, pp.109–14; and 15 January 1869, pp. 121–7

– 'The History of Alizarine and Allied Colouring Matters and Their Production from Coal Tar', JSA, 30 May 1879
– 'On Mauveine and Allied Colouring Matters', *Journal of the Chemical Society*, 1879, pp. 717–32
– Hofmann Memorial Lecture: 'The Origin of the Coal-Tar Colour Industry and the Contributions of Hofmann and His Pupils', *Journal of the Chemical Society*, Vol. 69, Part 1, 1896, pp. 596–637
– 'The Story of the Discovery of the First Aniline Dye', *Scientific American*, 10 November 1906
– 'The Magnetic Rotation of Hexatriene and Its Relationship to Benzene and other Aromatic Compounds, *Journal of the Chemical Society*, 1907, pp. 806–17
Perkin, William H. (Jnr), 'The Position of the Organic Chemical Industry', *Journal of the Chemical Society*, 25 March 1915
Phipson, Dr T. L., 'The Aniline Dyes', *Popular Science Review*, London, 1864, pp. 429–37
Punch, 7 and 21 August, 18 and 25 September, 13 and 20 November 1858; 16 April, 7, 14 and 28 May, 4 and 18 June, 16 July, 6 and 20 August, 3 December 1859
Rawson, Christopher, 'Valuation of Indigos', JSDC, 25 February 1885
Reed, Peter, 'The British Chemical Industry and the Indigo Trade', BJHS, No. 25, 1992, pp. 113-25
Reinhardt, Carsten, and Travis, Anthony S., 'The Introduction of Aniline to Paper Printing and Queen Victoria's Postage Stamps', *Ambix*, Vol. 44, March 1997
Robinson, Sir Robert, 'The Perkin Family of Organic Chemists', *Endeavour*, April 1956
Rowe, F. M., 'Tribute to Founder of Synthetic Dye Industry', *Dyer, Textile Printer and Finisher*, November 1938, p. 392
Schweitzer, Dr Hugo, 'The Influence and Effect of Perkin's Initiative', *The Dyer and Calico Printer*, 10 December 1906
Taylor, G. W., 'Identification of Dyes on Early William Morris Embroideries from Castle Howard', *Textile History*, 16, 1985, pp. 97–102
The Times, 3, 6, 9, 12 and 13 September 1884

Todd, Sir Alexander, 'The Future of Organic Chemistry', *Listener*, 17 May 1956

Travis, Anthony S., 'Science as Receptor of Technology: Paul Ehrlich and the Synthetic Dyestuffs Industry', *Science in Context 3*, 2, 1989, pp. 383–408

– 'Perkin's Mauve: Ancestor of the Organic Chemical Industry', *Technology and Culture*, Vol. 31, 1990, pp. 51–82

– 'Science's Powerful Companion: A. W. Hoffmann's Investigation of Aniline Red and its Derviatives', BJHS, 25, 1992, pp. 27–44

– 'August Wilhelm Hofmann', *Endeavour*, Vol. 16, No. 2, 1992, pp. 59–65

– and Benfey, Theodor, 'August Wilhelm Hofmann: A Centennial Tribute', *Education in Chemistry*, 1992, pp. 69–72

– 'The Man Who Put Science into Industry', *Chemistry and Industry*, 20 April 1992

– 'Poisoned Groundwater and Contaminated Soil: The Tribulations and Trial of the First Major Manufacturer of Aniline Dyes in Basel', *Environmental History*, 2, July 1997, pp. 343--65

Van Den Belt, Henk, 'Why Monopoly Failed: The Rise and Fall of Société La Fuchsine', BJHS, 25, 1992, pp. 45–63

Various authors, *The Athenaeum*, No. 1767, 7 September 1861

Various authors, 'Discussion on the Alleged Poisonous Action of Dyes on the Skin', JSDC, 25 November 1884

Various authors, 'Presentations to Sir William Perkin: Official Notices', *Journal of Science*, 31 August 1906

Various authors, 'Centenary Anniversary Tribute to Perkin', *Manchester Guardian*, 7 May 1956

Various authors, 'Sir William Henry Perkin', *Ciba Review*, No. 115, June 1956

Various authors, 'Proceedings of the Society', JSDC, Vol 72, No. 12, December 1956

Various authors, 'Dyes on Historical and Archaeological Textiles', 2nd Meeting, National Museum of Antiquities of Scotland, September 1983, pp. 14–16, 20–25

Vetterli, W. A., 'The History of Indigo', *Ciba Review*, 85, Basle, April 1951

Whittaker, C. M., 'Some Early Stages in the Renaissance of the British Dyemaking Industry', JSDC, Vol. 72, No. 12, 1956, pp. 557–63

Books

Anissimov, Myriam, *Primo Levi: Tragedy of an Optimist*, Aurum Press, London 1998

Arnold, David, *Imperial Medicine and Indigenous Societies*, Manchester University Press, 1988

Bedoukian, Paul Z, *Perfumery and Flavouring Synthetics*, Elsevier Publishing Company, 1967

Beer, John Joseph, *The Emergence of the German Dye Industry*, University of Illinois, 1959; new edition, 1981

Birren, Faber, *History of Color in Painting*, Litton Educational Publishing, Inc., 1965

Breward, Christopher, *The Culture of Fashion*, Manchester University Press, 1995

Brock, William H., *The Norton History of Chemistry*, W. W. Norton & Co., New York, 1993

Brockington, C. Fraser, *Public Health in the Nineteenth Century*, E. & S. Livingstone Ltd, Edinburgh and London, 1965

Bruce-Chwatt, Leonard Jan, and Zulueta, Julian de, *The Rise and Fall of Malaria in Europe*, Oxford University Press, 1980

Buck, Anne, *Victorian Costumes and Costume Accessories*, Herbert Jenkins, London, 1961

Bud, Robert and Roberts, Gerrylynn K., *Science Versus Practice*, Manchester University Press, 1984

Christy, Cuthbert, *Mosquitos and Malaria: A Summary of Knowledge on the Subject up to Date*, Sampson Low, Marston and Co. Ltd, London, 1900

Clements, Richard, *Modern Chemical Discoveries*, Routledge & Kegan Paul Ltd, London, 1954

Cumming, Valerie, *Royal Dress: The Image and the Reality 1580 to the Present Day*, B. T. Batsford Ltd, 1989
Cunnington, C. Willett, *The Perfect Lady*, Max Parrish & Co. Ltd, 1948
Desowitz, Robert, *The Malaria Capers*, W. W. Norton & Co. Ltd., London, 1991
Duran-Reynals, M. L., *The Fever Bark Tree*, W. H. Allen, London, 1947
Fenichell, Stephen, *Plastic: The Making of a Synthetic Century*, HarperBusiness, New York, 1996
Fox, M. R., *Dye-Makers of Great Britain 1856–1976*, Imperial Chemical Industries, Manchester, 1987.
Galdston, Iago, *Behind the Sulfa Drugs*, D. Appleton-Century Company, Inc., New York, 1943
Gernsheim, Alison, *Fashion and Reality*, Faber and Faber, London, 1963
Giles, Herbert M., and Warrell, David A., *Bruce-Chwatt's Essential Malariology*, Edward Arnold, 1992
Haber, L. F., *The Chemical Industry During the Nineteenth Century*, Clarendon Press, Oxford, 1958
Hall, G. K., *Du Pont and the International Chemical Industry*, Twayne Publishers, 1984
Harrow, Benjamin, *Eminent Chemists of Our Time*, T. Fisher Unwin Ltd, London, 1921
Hartnell, Norman, *Royal Courts of Fashion*, Cassell, London, 1971
Hayes, Peter, *Industry and Ideology: I. G. Farben in the Nazi* Era, Cambridge University Press, 1987
Heusser, Albert H., *The History of the Silk Dyeing Industry in the United States*, Silk Dyers' Association of America, New Jersey, 1927
Hills, Richard L., and Brock, W. H., eds., *Chemistry and the Chemical Industry in the 19th Century*, Variorum, 1997.
Hobson, W., *World Health and History*, John Wright & Sons Ltd., Bristol, 1963
Holden, Angus, *Elegant Modes in the Nineteenth Century from High Waist to Bustle*, George Allen & Unwin Ltd, 1935
Kendall, James, *Young Chemists and Great Discoveries*, G. Bell & Sons Ltd, London, 1939

Laver, James, *Taste and Fashion from the French Revolution to the Present Day*, George G. Harrap and Co. Ltd, 1945

Leggett, William F, *Ancient and Medieval Dyes*, Chemical Publishing Co., Brooklyn, NY, 1944

Matthews, Leslie G., *History of Pharmacy in Britain*, E. & S. Livingstone Ltd, Edinburgh and London, 1962

Megroz, R. L., *Ronald Ross: Discoverer and Creator*, George Allen and Unwin Ltd., London, 1931.

Meldola, Raphael, Green, Arthur G., and Cain, John Cannel, eds., *Jubilee of the Discovery of Mauve and the Foundation of the Coal-Tar Industry by Sir W. H. Perkin*, Perkin Memorial Committee, London, 1906

Moore, Doris Langley, *The Woman in Fashion*, B. T. Batsford Ltd, 1949

Morgan, Sir Gilbert T, and Pratt, David Doig, *British Chemical Industry: Its Rise and Development*, Edward Arnold and Co., 1938

Morris, Peter J. T., and Russell, Colin A., *Archives of the British Chemical Industry 1750–1914*, British Society for the History of Science, Oxfordshire, 1988

Nye, Edwin R., and Gibson, Mary E., *Ronald Ross: Malariologist and Polymath*, London, 1997

Pelling, Margaret, *Cholera, Fever and English Medicine 1825–1865*, Oxford University Press, 1978

Perkin, Arthur George, and Everest, Arthur Ernest, *The Natural Organic Colouring Matters*, Longmans, Green and Co., London, 1918

Pilcher, Richard B., and Butler-Jones, Frank, *What Industry Owes to Chemical Science*, Constable and Co. Ltd, London, 1918

Ponting, Ken, *A Dictionary of Dyes and Dyeing*, Bell and Hyman Ltd, London, 1981

Portugal, Franklin H., and Cohen, Jack S., *A Century of DNA*, MIT Press, 1977

Ramsey, Albert R. J., and Weston, H. Claude, *Artificial Dyestuffs: Their Nature, Manufacture and Uses*, George Routledge & Sons, London, 1917

Reinhardt, Carsten, and Travis, Anthony S., *Heinrich Caro and the Creation of Modern Chemical Industry*, Kluwer Academic Publishers, The Netherlands, 2000

Rhys, Grace, *Modes & Manners of the Nineteenth Century*, J. M. Dent & Co., London, 1909

Ross, Ronald, *Memoirs*, John Murray, London, 1923

Routledge Robert, *Discoveries and Inventions of the Nineteenth Century*, George Routledge and Sons, Ltd, London, 1900

Rowe, Frederick Maurice, *The Development of the Chemistry of Commercial Synthetic Dyes (1856–1938)*, The Institute of Chemistry of Great Britain and Ireland, 1938

Scott, H. Harold, *A History of Tropical Medicine*, Edward Arnold & Co., London, 1938

Singer, Charles, *The Earliest Chemical Industry: An Essay in the Historical Relations of Economics and Technology Illustrated from the Alum Trade*, The Folio Society, London, 1948

Spanier, Ehud, ed., *The Royal Purple and the Biblical Blue: The Study of Chief Rabbi Dr Isaac Herzog on the Dye Industries in Ancient Israel and Recent Scientific Contributions*, Keter, Jerusalem, 1987

Stephens, J. W. W., *Blackwater Fever*, Hodder & Stoughton Ltd., 1937

Taylor, Lou, *Mourning Dress: A Costume and Social History*, George Allen & Unwin, London, 1983

Taylor, Norman, *Cinchona in Java*, Greenberg, New York, 1945

– *Plant Drugs that Changed the World*, George Allen & Unwin Ltd., London, 1966

Tordoff, Maurice, *The Servant of Colour*, The Society of Dyers and Colourists, Bradford, 1984

Tozer, Jane, and Levitt, Sarah, *Fabric of Society*, Laura Ashley Ltd, Wales, 1983

Travis, Anthony S., *The Colour Chemists*, Brent Schools and Industry Project, 1983

– *The Rainbow Makers: The Origins of the Synthetic Dyestuffs Industry in Western Europe*, Lehigh University Press, Bethlehem; Associated University Presses, London and Toronto, 1993

Various authors, *Proceedings of the Celebration of the Three Hundredth Anniversary of the First Recognised Use of Cinchona*, St Louis, USA, 1931

Various authors, *The Life and Work of Professor William Henry Perkin (Jnr)*, The Chemical Society, London, 1932

Various authors, *Perkin Centenary, London: 100 Years of Synthetic Dyestuffs*, Pergamon Press, 1958

Various authors, *Malaria: Parasites, Transmission and Treatment*, The London Cinchona Bureau, 1963

Various authors, *Proceedings of the Second Annual Conference of the Costume Society: High Victorian Costume 1860-1890*, London, 1969

Venkataraman, K., *The Chemistry of Synthetic Dyes, Organic and Biological Chemistry: A Series of Monographs*, Academic Press, New York, 1952

– *The Analytical Chemistry of Synthetic Dyes*, John Wiley & Sons, 1977

Waddington, Herbert, *The Story of a Family Business Expanded and Revived*, Leeds, 1953

Wagner, Rudolf Von, *Manual of Chemical Technology*, translated and edited by William Crookes, J. & A. Churchill, London, 1892

Walkey, Christina, *The Way to Wear 'Em: 150 Years of Punch On Fashion*, Peter Owen, London, 1985

Watts, Henry, *A Dictionary of Chemistry, and the Allied Branches of Other Sciences*, Longmans, Green and Co., London, 1870

Waugh, Nora, *The Cut of Women's Clothes 1600–1930*, Theatre Arts Books/Methuen, New York, 1968

Weatherall, M., *In Search of a Cure*, Oxford University Press, 1990

White, Howard J., ed., *Proceedings of the Perkin Centennial*, American Association of Textile Chemists and Colorists, New York, 1956

Whittaker, C. M., and Wilcock, C. C., *Dyeing with Coal-Tar Dyestuffs: The Principles Involved and the Methods Employed*, Baillière, Tindall and Cox, London, 1949

Wingate, P. J., *The Colorful Du Pont Company*, Serendipity Press, 1982

INDEX

INDEX